ILS_ INNOVATIVE LEARNING SPACE

MASSIMO FAIFERRI
SAMANTA BARTOCCI
FABRIZIO PUSCEDDU

SARDEGNA RICERCHE

This publication collects the results of the scientific school ILS_Innovative Learning Spaces organized from 27 August to 3 September 2016 in Porto Conte Ricerche with the contribution of Sardegna Ricerche.

INDEX

ILS FOREWORD

The reasons behind a school

Massimo Faiferri*

The ILS Innovative Learning Spaces Scientific School is run by Sassari University's Department of Architecture, Design & Urban Planning (DADU) as part of the SPIN-APP Innovative Spaces for Learning project (Regional Law 7, *Promotion of scientific research and technological innovation in Sardinia)*, in collaboration with the National Nuclear Physics Institute (INFN), and the contribution of the Sardegna Ricerche project *Scientific School 2015*.

The school addresses students, undergraduates, graduates or young professionals who wish to embark on a course with high-quality scientific merit on the theme of innovative learning spaces, paying particular attention to space design, ICT aspects and energy sustainability. The school envisages the admission of a maximum 35 students, encouraging local participation but also diversity in educational paths and a cultural exchange between students of different origins and backgrounds in the field of architectural and urban planning, design, informatics, pedagogy, materials science and energy control, recovery and production.

The general objective of the ILS Scientific School is to study (in the form of Summer School workshops) innovative design for learning environments, that may be used as a reference base and model for constructing or adapting school buildings. Getting involved in design for the school environment means to responsibly reappraise the role of space in the dynamics of learning processes, as well as think about renewing the way teaching is organised. The prerequisite is the creation of a flexible, modular, multifunctional space fit for the requirements of the students' individual paths as they develop skills, which will incorporate principles and materials inherent in good, sustainable practices as its constitutive elements and accommodate new technological instruments as an enhancement of real space. The most recent European funding programmes for research (Horizon 2020) recognise education as a key element for the growth of human capital able to cope with the challenges of an inclusive, innovative and safe society, emphasising the role of the school and the learning processes underpinning policies of social development, cohesion and wellbeing.

In the objectives of the common strategic framework 2014-2020 the European Commission puts the development and implementation of new ideas (products, services and models) able to respond to social needs and expectations in the 'social innovation' category. It specifies in particular that technological and social innovations – in this case incorporated in 'school design' – can use the potential of the networks and ICT to strengthen the autonomy of individuals and for social inclusion, starting with the primary school. The analysis of the physical environments of learning from a didactic point of view has to contend with the digital revolution in schools and leads us to review interaction, in particular,

* Massimo Faiferri, DADU, University of Sassari.

between the organisation of the physical environment in which structured learning is fostered (schools) and participation in virtual environments that are the result of information and communication digital technologies.
These issues are at the centre of attention in programmes of international institutions like the OCSE, with the Centre for Effective Learning Environments (CELE) (which published the book *Designing for Education: Compendium of Exemplary Educational Facilities* in 2011) and the Innovative Learning Environment (ILE) project. Much research underway concerns the theme of teaching and learning spaces encouraged by guidance of a European and national kind, but also at an international scale, such as the studies developed under the Microsoft Innovative Schools Programme project or the AltSchool, founded in 2013 by one of the former coordinators of Google in Silicon Valley, which recently received funding to the tune of 133 million dollars from one of the main world investment funds. In Italy the MIUR started up a series of initiatives at the time under Minister Profumo in support of reappraising school spaces, summarised in the *Linee Guida per l'edilizia scolastica per scuole più sicure e spazi di apprendimento al passo con l'innovazione digitale (Guidelines for building safer schools and learning spaces abreast with digital innovation)*, while the construction of novel scenarios for the school of the future has been the main objective of the National Institute of Documentation, Innovation & Educational Research (INDIRE) and various groups belonging to university research platforms, partially summarised in the "Abitare" initiative *Re-inventing the school*. Most of the cases were studies that were too sectoral, however, in analysing the problem and of potential project-based developments, as they were either handled with a single-subject approach or the experiments were not implemented on the territory in terms of comparison and feasibility.
According to the Chairperson of the Foundation on Economic Trends, the crisis of traditional economy and finance, together with the emergence of the global warming issue, are a chance to turn a crisis into opportunity and bring about an authentic revolution in the way of producing, storing and distributing energy. In this perspective the focus is on buildings, especially public ones and those with important functions, like schools, to be reconverted into 'power plants' once they have been equipped with instruments able to optimise energy efficiency. The aim is to gradually construct urban nuclei and whole cities that are self-sufficient from the energy viewpoint, able to produce and accumulate power thanks also to the appropriate use of the resources available in daily activities. This objective may be made possible on the basis of the integration into existing or new school architecture, of non-invasive, low-cost devices that will use traditional power resources with a high yield but also be able to transform the energy available in nature into electricity.
All this requires a replanning process of the buildings that will incorporate instruments integrating renewable energy sources and limiting peak loads, forms of collective and participative contribution in the global electricity market with a Smart Cities vision in line with Italian and European normative standards, which envisage energy certification of buildings, favouring recourse to renewable energy sources.
Planning for spaces of education has to contend with a series of themes with various meanings that permeate teaching and learning in the contemporary society, such as the adoption by schools of new communication technologies, new cooperative learning methods, interdisciplinarity, the quest for new expressivity and languages, and reciprocity of learning and teaching. The introduction of ICT as a didactic tool is able to bring about a drastic change in ways of acquiring, producing and passing on knowledge; at the same time, communicative relations have changed, in that passive learning may

be completely surpassed by more dialogical learning. With the use of digital technologies, lessons can become interactive and students can be the principal actors, acquiring the information they need from a variety of sources and organising it with the help of their classmates and teacher.
The analysis and planning of the physical environments of learning from a didactic point of view have to contend with the digital revolution in schools and lead us to review interaction, in particular, between the organisation of the physical environment in which structured learning is fostered and progresses (schools) and the potential of interaction and participation in virtual environments that are possible and experimented with thanks to information and communication digital technologies. At the same time, as regards the educational proposals and energy requirements linked with the new technologies, schools need to aim at self-sufficiency in energy, not just for ethical questions but from the point of view of management and functionality, seeking to be self-sustaining by using eco-compatible materials. Planning has to consider forms and methods linked with increasingly efficient use of the power resources available, but must also explore ways of generating electricity in a 'cleaner' manner and recuperating (when possible) unused power or power in natural phenomena that is wasted.
The ILS Scientific School has organised various activities – lessons, seminars, conferences and workshops - in which small groups of students (each under the guidance of at least one tutor or lecturer) have worked on a proposal involving either the use and integration of ICT instruments in learning environments, or forms and methods of production/recovery of energy in school buildings.
The panel of internationally renowned lecturers includes well-known architects, experts in planning contemporary schools and lecturers at prestigious universities with experience on the topics studied, as well as researchers in the field of energy and applied ICT. The structure of the ILS Scientific School is founded on the principles of the Ecourbanlab research workshop, which pursues research methodology based on developing relations of both an interdisciplinary type and between theory and practice components, investing research experience in teaching and dealing with the context so as not to lose contact with the real world.
Ecourbanlab has developed its research work and didactics on the basis of 'pretexts', in such a way that it is the project for space in the physical context that has provided an anchorage for the theoretical hypotheses which, one after another, have been explored in relation to the issues handled. This attitude implies a close bond with the territory and its social and administrative components: what the design process aspires to is not to interpret an actual world but to define a possible reality, consisting not of properties but of relations between things.
On these foundations the ILS Innovative Learning Spaces experience was born, the international summer school that has working groups of young planners, students and professionals confront each other in interdisciplinary working teams under the guidance of renowned international architects: the results of the first edition of the ILS Scientific School are collected in this publication.

Space design for learning

Massimo Faiferri*

In the scientific field it is recognised nowadays that the development of cognitive capacities takes place through physical movement, in particular through experiments that we are able to devise as children through body movement. The continuous perceptive stimuli that accompany growth, like (for example) the action necessary to grasp, or changes of position for balance, enable the child to initiate the process of acquisition of knowledge of the world surrounding him and gradual awareness of his own body and therefore himself.

The discovery of mirror neurons – so-called because in the neural system of a person witnessing a certain behaviour in another, they cause a specular reaction to that of the action observed – constituted a basic turning-point to explain how we project ourselves emotionally in the actions of others or how we enter into empathic relations with the forms of the built environment. The conclusions of the research group coordinated by Giacomo Rizzolatti show that we imitate other people's actions (often unawares or precognitively) in the same areas of the brain that we use for our actions. This observation, corroborated experimentally in 2010 by Marco Iacoboni, confirms that seeing an object involves automatically recalling the action we would carry out with it, just as observing someone's behaviour involves simulating the corresponding potential action. Generally speaking, mirror neurons therefore appear to phenomenologically define our empathic relationship with the world: the body has a fundamental role in constructing the mental picture of each of us – a picture that will change continuously during life, making us different depending on what we do, what we learn and what happens to us every day.

This process is more than obvious in the case of motor disciplines. By observing the movements of athletes' bodies in hitting or kicking a ball, children learn very quickly to imitate their movements since, as is now known, their neural circuits simulate or repeat the athletes' muscular movements. The mirror neuron system is essential also for understanding a series of other actions, such as how we form our sense of self, how we identify ourselves and relate socially, how we are able to have spatial awareness and also how we develop language. We look at, touch, listen to and assess the world with all our bodily existence, organising and defining the perceivable world around our body. The empathic system of mirror neurons is not restricted to the sense of sight, for it is through all the senses that we constantly understand and learn, building up our sense of belonging to a space, a place, a community. As we grow, we acquire physical-body skills not so much taught through language as by the direct transfer of a muscular skill from teacher to learner through an act of sensitive perception and bodily imitation.

In one of his most interesting studies entitled *Mi ritorno in mente. Il corpo, le emozioni, la coscienza* (I come back to mind. Body, emotions, mind) Edoardo Boncinelli writes about the body, emotions and mind, trying to focus on the relationships between these entities to trace back their links through

* Massimo Faiferri, DADU, University of Sassari.

the problematic node of rationality and its relationship with the mind. Constantly in touch with the most advanced neuroscience theories, Boncinelli tries to delineate a theory of 'I' in the last chapter of the book, as well as its relationship with the world and the body, in these terms: *"perception is always aimed at action, but there cannot be action without motivation or positive expectation. Perception and the cognitive mind suggest 'how' we should act: emotivity gives us a reason for acting and urges us to do so"*[1].

Action is therefore the essential element for achieving any learning. The relation between single actions or more complex action patterns creates the conditions for intelligently organising action, on the basis of experience that may be transferred to other situations or the organisation of new actions. Maria Montessori herself noted that the simple observation of a child highlighted that the development of its mind passed through movements, revealing how deeply the mind is 'incorporated' and 'anchored' in the body. It is a matter of synchronicity between acting, thinking and speaking, which challenges the classic idea of incoming sensory information being processed via a linear course that ends with the production of a consequent motor outcome. Action, in effect, is not the final result of the perceptive process but an integral part of this process and is inseparable from the sensory stimulus, as it is contained in it. Every possibility of access to knowledge comes from the body, therefore, and body action is indispensable for the development of cognitive processes.

The tactile body

Right from the early days of life, experiencing the world means above all to experiment with the key role of the brain-body pair in the relationship with buildings; for this to develop, an architectural space conceived as an emotional and multisensorial experience is necessary. Unfortunately, the most widespread pedagogies and educational practices continue to keep mental, intellectual and emotional capacities separate from the senses and multiple dimensions of corporality. The body is taken into consideration in sport and dance, just as the senses are acknowledged in direct connection with art and musical education, but seldom is bodily existence recognised as the foundation of awareness and understanding of self. Juhani Pallasmaa is one of the main architects to have studied these concepts in depth; in *The Eyes of the Skin: Architecture and the Senses*[2], one of his best known and influential contributions, he maintains that most contemporary schools, like most buildings, do not manage to activate any of our senses, except sight, thus remaining alienated from the primary process of knowledge. According to Pallasmaa, architecture has turned into a visual art that pursues immediately comprehensible retinal images based on the predominance of focused vision and is largely imagined and realised within the exclusive sense of sight. The result of this approach cannot but be a levelling out of the potential of architectural buildings and the environment, reduced to a stage set devoid of material, constructive authenticity. In direct contrast with this conception, Pallasmaa's text develops a series of arguments seeking to claim a role for the other senses – in particular, touch – in the experience and understanding of the world; this text considers all the senses, including sight, as extensions of touch, namely different ways of 'touching' the world, as a skill of the epidermal tissue. Together with

---

[1]. Edoardo Bonincelli, *Mi ritorno in mente. Il corpo, le emozioni, la coscienza,* Longanesi, Milan 2010, p. 81-82.
[2]. Juhani Pallasmaa, *The Eyes of the Skin: Architecture and the Senses,* John Wiley & Sons, Chichester 1996, Ed.1, Italian trans. *Gli occhi della pelle. L'architettura e i sensi,* Jaca Book, Milan 2007.

unfocused peripheral vision, tactility becomes the most authentic essence of experience had, as the author explains: *"Whereas focused vision sets us against the world, peripheral vision envelops us in the fibres of the world. Besides criticising the hegemony of vision, we need to reconsider the essence itself of sight"*. With these presuppositions it is indispensable to think of architectures that envisage multi-sensory experiences, creating, that is, the co-presence of sensations able to bring the entire perception of our body into a relationship with the built environment: *"An architectural work is not experienced as though it were a series of isolated retinal pictures, but in the total integration of its material, bodily and spiritual essence"*. Pallasmaa refers here to the multisensory perception of architectural space, which binds together the structure and sound of the space, the temperature, material grain of the surfaces, the objects found in it, the tension between inside and outside, the intimacy and, last but not least, the light, which Peter Zumthor abbreviates using the term 'atmosphere'.

In general, we may state that we live in world that is prevalently visual: the tactile part of vision has almost disappeared. One of the many reasons for this is that the visual world has an instantaneous quality, while the tactile world is very slow: the visual world is public, whereas the tactile world is always private. The following are a couple of the various examples Pallasmaa gives to explain this substantial difference. The first concerns the description of *La finestra aperta*[3] by Matisse, which, in his opinion, cannot be reduced to the simple rational picture of an interior, with a balcony opening onto a harbour, because our brain begins to associate this image with the damp heat of the room, the sounds of the harbour and the perfume of the plants adorning the balcony. Similarly, the memory of our years of studying in the past cannot be limited to a simple reconstruction of an image, but is enhanced and implemented through 'sensorial kingdoms', which contribute to defining our experience. *"I still remember the smell of my first day at school 65 years ago.* – Pallasmaa wrote on the subject in 2010 - *It was during the war years; I had been taken by my grandfather from the town where we lived into the Finnish countryside. I went to a rather simple school, but in some ways it was an advanced education. In those times, the students of an agricultural community took turns going to school at 6.30 in the morning, before the other students arrived at 8.30, to light the fire in the stove to heat the school. Looking back, I appreciate those dark mornings in the cold school building. It created the feeling that the school was not just an abstract place for learning; it was integral part of life. The smell of burnt wood and the soap used to wash the wood flooring: those smells are for me the essence of the school"*[4]. Tastes and smells can be points of reference, orientation and memory; we feel better and improve our learning capacities when we are strongly rooted in our life experiences. Creating the spatial and sensorial conditions to root children in real life therefore becomes indispensable; rooting means safety and safety favours the mental space for learning; children of all ages need places where they can learn by touching, manipulating and using things with their hands. Consistent with the recent discoveries in the field of neuroscience which, as stated, show correlation between motor patterns and cognitive functions, it becomes necessary to conceive architectures in which spaces are able to provide those using them with opportunities for

[3]. Henri Matisse, *Fenêtre ouverte à Collioure,* 1905, oil on canvas, John Hay Whitney Collection, National Gallery of Art, Washington. The work was created during a stay at Collioure, on the southern coast of France: Matisse painted the panorama of the harbour and boats on the sea from the window of his hotel room.

[4]. Juhani Pallasmaa, *Allowing Children to be Rooted in Real Life,* in Various Authors, *The Third Teacher,* Abrams, New York 2010, p. 176.

action, enabling them to grasp how that same space will be in the future. As part of this process, the senses are the primary instrument to develop learning linked with the complex sequence of time-space relations and movements in carrying out a task. A way to activate the senses has to be found, through projects able to manage sounds, smells, tastes, tactility and memory of movement: environments full of sensory experience, that will help students to keep and retrieve what they learn. *"Unfortunately, the predominant educational philosophies continue to emphasise and valorise conceptual, intellectual and verbal knowledge over and above the silent, non-conceptualised wisdom of our embodied processes.* - explains Pallasmaa - *Nowadays this point of view persists in spite of all the overwhelming evidence of this catastrophic misunderstanding, provided by philosophical arguments and recent developments and discoveries in neurology and the cognitive sciences"*[5].

Architectures of relations

Architectural design, especially that for learning spaces, must therefore be – today more than ever – the design of relations rather than objects, of dynamic spaces rather than static scenes, since it is increasingly necessary to draw attention to the relationships between things and people. The project for a school must build a system of relations, where space becomes a precious instrument to define and organise educational relations. This process is always contiguous with landscape design, with its generative and evolutive processes; landscape design cannot, in effect, yield a static, unchanging image, whereas formalism has to give way to a complex metabolism of functioning, arising from the relationship between man and territory. These observations underpinned João Nunes' participation in the 2016 ILS Summer School, also as regards his way of understanding the landscape: *"The landscape has, over the course of time, changed its meaning to the point of being translated nowadays into the set of footprints left on the territory by the various communities and individuals that have shared it, superimposed on those of the physical genesis of the actual territory and those of the transformations from which the living community is alienated. In short, it is a set of footprints coded according to a system of meanings; the landscape will therefore be the whole set of relations these footprints match as perceptible manifestations of life: relations that develop between individuals of the same community, between individuals of different communities, between different communities, collectively, and between all these and the territory; relations that entail an effort to survive, a mechanism to ensure survival of the community, a gesture of protection by preceding generations of the following ones"*[6].
At the same time, architectural design that aims to develop through a similar system of relations should be structured by similar methods, e.g. by creating a dynamic space, full of internal tensions, emotionally sensitive and able to introduce the concepts of ambulatory and peripatetic space for experimentation through the senses, referring back to experiences of the picturesque but also to the phenomenological concept of space based on movement, as is found in some of Steven Holl's work. In 1995, calling his competition project for the new contemporary art museum of Helsinki *Kiasma,* Steven Holl in some way disclosed one of the basic principles of his way of conceiving and devel-

---

[5]. Juhani Pallasmaa, *La mano che pensa. Saggezza esistenziale e incarnata nell'Architettura*, Safarà, Pordenone 2014, p. 23, Italian trans. of *The Thinking Hand: Existential and Embodied Wisdom in Architecture*, John Wiley & Sons, Chichester 2009.

[6]. João Nunes, *Paesaggi, passaggi,* in "Lettera internazionale", 105, 2010.

oping architectural planning. As he himself admitted, the *Kiasma* project explicitly took up Maurice Merleau-Ponty's phenomenology of perception and concentrated on the theme of perception, understood not so much as a pure glance, able to describe phenomenal structures and essences, as, rather, man's primordial experience. Far from unambiguous stylistic and formal research, Holl's idea of architecture is based on realising spaces with complex perception, featuring a refined, varied use of materials and light, able to let users experiment with a multisensory perceptive experience: *"It is not the dimensions of things that creates a space,* Holl stated – *but it is a quality connected with perception: only through the latter do we create a space"*. Through a building, by concentrating on the relationship between body, space, eye and mind, it becomes the instrument of understanding architecture, allowing it to become alive and show itself in all its complexity, giving it the possibility to change, also and above all by means of the light, which will never be the same. Very briefly, the concept of space in Holl's architectures may be summarised as a cross between these experiences, apparently separate, that man alone, with sensory experience, is able to inhabit, measure, understand and evaluate. As he himself wrote: *"My battle is to find and extract the phenomenal potential of the basic idea. I believe that the proof of truth in architecture lies in the phenomena generated by a body that moves in space: a thing that can be felt and perceived even without understanding the programmatic intent of the architect. In this sense architecture is a universal language"*.
More and more contemporary art museums are considered places of interaction between art and users rather than simple display places. This attitude is clear in *Kiasma,* where public participation in the activities of the museum, in particular by children, is greatly encouraged: *the Helsinki museum,* Holl writes, *shows how the body is the true gauge of superimposed space. The cross between the concept of building and the interweaving of the landscape, light and city indicates several routes through the museum, entailing body movements and parallax of the spaces that open up. The body becomes a live spatial gauge, a moving one, through the extended and superimposed perspective"*[7].
The experience of architecture thus becomes much more involving than any other visual experience, including the cinema; at the same time, it implicitly manifests the importance of the theme of the pathway for planning. The history of architecture proposes various interpretations of this compositive instrument, supplying countless examples that are based on and develop from the pathway, to mention just one, Le Corbusier's *promenade architecturale*. But the sense the pathway could take on as a compositive principle in the research context under discussion is something different. On the one hand, the 'architectural promenade' is one of the essential themes of Le Corbusier's work (taken up to a different extent in the works of other modern masters and their epigones, just as is clear also in numerous works of the past, characterising, let us say, a good period of the history of architecture); on the other, this principle appears as a possible key to interpret the relationship between space and learning.
The arguments put forward by another stimulating book that develops an educated reflection on the influence of walking in the perception of architecture, enable the importance to be grasped of this principle in designing a building or space destined for learning. We are talking of *Motion,* émotions.

---

[7]. Steven Holl, *Parallax. Architettura e percezione,* Postmedia Books, Milan 2004, p. 19, Italian trans. of *Parallax,* Princeton Architectural Press, New York 2000.

Architettura, movimento e percezione (Motion, emotions. Architecture, movement and perception), a collection of papers by Jacques Gubler[8] who, by analysing and comparing different experiences in the history of architecture, attempts to study the complex relationship between space, movement and perception. The first essay of *Motion*, émotions, *Note sulla camminata e l'architettura del suolo (Notes on walking and ground architecture)* explores very briefly the indivisible relationship between movement and learning. Centred on the relation between perceptive mechanisms and architecture, through a re-reading of some philosophical (from Kant to Rousseau) and architectural (from Le Corbusier to Lloyd Wright) extracts, Gubler's text implicitly demonstrates how learning is strictly connected and correlated with action. A quotation from Rousseau's educational syllabus should be enough, illustrated in Émile *ou De l'education*, where we read: *"we will turn every feeling into an idea. The animal and wild virtues of the human machine are the first experience of the world: our first masters of philosophy are our feet, our hands, our eyes"*, to highlight that the route on foot is the starting-point of geographical knowledge.

"It is only on the strength of walking, touching, counting and measuring that we learn to estimate dimensions [...] Natural gauges exist that are more or less the same everywhere: man's paces, his extended arm, his height. When a child estimates the height of a level plane, his educator may be of use to him as a unit of measure; if he estimates the height of a bell-tower, the unit will be the houses. If he wants to know how many leagues he has covered, he can count the hours of walking."

But it is also by re-reading the theories of some of the masters of modern architecture that Gubler reinforces our convictions. For example, read the extract from Wright's autobiography dedicated to childhood memories: the events of the young Frank Lloyd, seen through walks with his Uncle John, are described as a true preamble to his learning of the world and architecture. With this reference, Gubler associates Wright's autobiography with the 'coming-of-age' novel genre and considers it a treatise of didactics and architecture. *"The ideal and the utopia, the democratic idea for Sullivan, organic architecture for Wright, are presented as arrival points of adventurous lives. The initial theme of the walk is subordinated to the discovery of agriculture, or the intact earth, sometimes forest, sometimes desert"*[9]. In Gubler's pages, the pieces taken from Lloyd Wright's or Sullivan's autobiographies refer to the literary theme of learning in the world of two boys, through the experiences of their early years in direct contact with nature and in symbiosis with the changing seasons – an educational experience of infancy/adolescence that will evolve over time, to become the basis of their urban theories.

Art and senses: thinking with the body

Design that intends to develop through a network of relations, focusing mainly on those generated by movement of the body in space and the sensory stimuli that space is able to create, cannot escape comparison with experiences developed in similar spheres. Renato Bocchi's research on the *relationship between art architecture space and landscape*[10] enables, for example, some artists' work founded on design for relations to be studied in depth. One of these is Giuseppe Penone, who contests with his

[8]. Jacques Gubler, *Motion*, émotions, Infolio, Gollion 2003, Italian trans. *Motion*, émotions. *Architettura, movimento e percezione*, Christian Marinotti, Milan 2014.

[9]. Op. cit., p. 22.

[10]. Renato Bocchi, *Progettare lo spazio e il movimento*, Gangemi, Rome 2009.

creations the current vision of rational man and culture, in contrast with nature and its wild character. Penone's works suggest an interpretation that is diametrically opposite: man is considered an integral part of nature, in a direct relationship with the elements that animate it. This interpretation, far from some of the 'environmentalist' positions reigning, suggests, rather, a direct relationship with nature, almost in symbiosis. What interests us here of this approach is the exceptional importance given to sensoriality as an instrument of knowledge and artistic expression, necessary for the development of the idea of identification between man and nature. It is a changing process that varies with the passing of time and brings Penone's artistic research close to an idea of architecture able to embody the temporal and procedural dimension of design. It is through the senses, therefore through the body, that man can get close to nature and penetrate it, to the point of knowing its most intimate vital processes; interacting with them, one learns how to understand them and be part of them. *"The breadth of a work of art will always be sense-friendly.* – writes Penone – *Observing, touching, feeling, trampling man's actions accumulated on the ground. A work of art is based on the senses and the logic that arises from them. I sense the tree swell round my hand resting on its trunk. Diapason tree; my ear resting on a tree trunk to hear its years of growth, to hear the sound of the wind swishing through its branches, its trunk, its roots deep into the earth. Each species of tree a sound, each day of the tree a different sound"*[11].
The work of another artist, Alberto Carneiro, also proposes a similar interpretation; in his works the identification – if not authentic symbiosis – is discovered between man and nature. Carneiro's works give importance to the action of the body over matter, via a process of appropriation-manipulation of nature and consequent transformation. Through corporality and sensuality, the body takes on the role of go-between for knowing the world, as the artist explains: *"Between my body and the ground a profound affinity has always existed. The wood and mountain I work with in a tree trunk or stone are an integral part of my being. My work is always a totalising appropriation of matter, recreated at two levels: that of fierce possession through the existential fury of the senses and that of mental possession for the need to be reunited with my roots"*[12].

Space and pedagogy

Learning environments should therefore be conceived on the basis of the set of notions we have examined, reappraising the role space can cover in learning processes, apart from renewing didactic organisation. Pallasmaa is very critical – as has been seen – of the school buildings of the present, on which he reports both the difficulty of interaction with the multi-sensoriality of the body, and the impossibility of learning from the life unfolding therein. Whereas school space should constitute the space of experience and therefore be active towards the children. The senses are the privileged ports of access for discovering the world and only by exercising them is it possible to fine-tune them, so much so that children continuously experiment with their environment, linking up sounds, smells, colours, tastes, temperatures and consistencies.

---

[11]. Giuseppe Penone, *Respirare l'ombra, Respirar la sombra,* Xunta de Galicia, Santiago de Compostela 1999.
[12]. Alberto Carneiro, *Notas para un manifesto de arte ecologica,* in *Alberto Carneiro,* Centro Galego de Arte Contemporanea, Santiago de Compostela 2001, exhibition catalogue.

A school building planned so that it will be able to develop and implement these processes cannot be conceived only in normative or construction terms. It goes without saying that good construction must underpin planning for a school, as for any building, just as respecting regulations (complex, above all in the Italian context) is a mandatory commitment for designing a public building; nevertheless, a careful analysis of the state of school buildings in Italy shows that these expectations have repeatedly been ignored. The decay of school buildings (outcome of a policy that has never invested in the sector) is a symptom of a sick Country, unable to meet the challenge of the State school as a great social infrastructure, with a capillary, open nature, as well as welcoming after teaching hours. To reverse the process is as complex as it is now inescapable: it is time to begin taking care of society, not just by making buildings safe but launching wider-reaching, more ambitious action, investing in the future generations. Already in 1947, Ernesto N. Rogers[13] pointed out in a "Domus" editorial how public expenditure cuts in Italy had distanced school construction from the field of architectural experimentation, to confine it to the sphere of bureaucratic competences of the technical apparatuses, downgrading its pedagogical potential and standardising school buildings within the realms of everyday prefabrication. The consequence of this was to deny the architectural discipline a chance to contribute to constructing a physical reality able to assist the development and sharing of knowledge, that could not solely be limited to pedagogical aspects. Educating architecture – like that Rogers wanted – needs an authentic revolution in social values and an investment policy at all levels of education, capable of considering architectural space a genuine pedagogical instrument.

In the first post-war period reflections on school building design were developing in which the result of the new pedagogical connotations was spreading parallel to the spatial qualities of the buildings, promoting together both a renewed interest in the individual and the need to coherently redefine spaces and places of instruction. In the experimental teaching methods of those years the child was no longer a passive subject in the hands of the educator but independent and active, the addressee of the attention and stimuli necessary – coming also from the educational environment – for individual development. The requirements of active didactic methods, based on encouraging capacities of observation, socialisation and movement, as well as the harmonic mental and physical development of the child, encountered the experimental needs of modern architecture. Convergence thus emerged on the necessity to break up the inflexibility of the typical nineteenth-century set-up of classrooms and central corridor, conforming with the inflexibility of the educational system, to experiment with new types of set-up, in contact with nature and equipped with collective environments and spaces for both special and open-air activities.

The recent history of Italy has shown how this desire for research and experimentation developed over time through some of the most advanced pedagogical approaches of the twentieth century: from Maria Montessori and Mario Lodi, through the early discussions on the 'school as home' developed by the lesser known Carolina and Rosa Agazzi, up until the important experiences of *Reggio Children,* which came to maturity with the extraordinary work – internationally famous – of Loris Malaguzzi. This was a range of important pedagogical reflections, now historic, belonging to the innovation process inherent in the contemporary way of considering the school, though not yet in how to teach or be at school and, above all, to physically and spatially design the school.

---

[13]. Ernesto Nathan Rogers, *Architettura educatrice,* in "Domus", 220, June 1947.

The matter of school space in Italy can immediately be grasped from the panels of the exhibition organised by Massimo Ferrari and Claudia Tinazzi for the Milan Triennial (set up in Sardinia, too, as part of our research path[14]); it has gone through ups and downs between elegant, cultured experiments and stalling phases in research on types of school building, partially interrupted (though at intervals of a decade) by the creation of significant buildings. At the same time, the exhibition shows that the theme of education, translated architecturally into scholastic institutes at all levels, is still one of the central issues of the contemporary period, as much from a political as a civic viewpoint: it is easily understood how, during the course of last century, architecture for schools marked out, through thick and thin, the history of the Country. There were numerous projects for scholastic facilities that represented important parts of the history of Italian architecture: the experiments of the 5th Milan Triennial (1933) and the works of Terragni (St. Elia nursery school, 1936), Cattaneo (Garbagnati nursery school, 1935-37), Giò Ponti (Bruzzano nursery school, 1934), Figini and Pollini (Olivetti crèche in Ivrea, 1939-41), characterising the first half of the last century, followed by those designed during the reconstruction years by Giuseppe Vaccaro (nursery school in Piacenza, 1953-62), Luciano Baldessari (nursery school in Milan, 1959-60), Mario Ridolfi (Olivetti day nursery in Ivrea, 1953-62), Ludovico Quaroni (Olivetti middle school in Ivrea, 1959-64), and those designed approximately a decade later by Alberto Samonà (national competition for ideas for compulsory schools in Bologna, 1964), Vittorio De Feo (schools in Rome, 1967, and Terni, 1968), Cappai and Mainardis (school in Oriago, 1968), to come a decade later to Guido Canella (various schools in Lombardy, 1968-75), Carlo Aymonino (campus in Pesaro, 1975-81) and Aldo Rossi (schools in Fagnano Olona, 1972-76, and Broni, 1979-81).
Sandy Attia and Beate Weyland, lecturers at our Summer School, are trying to understand through their research how the important Italian pedagogical experiments of the past could harmonise with the extraordinary tradition of school construction that emerged during the Triennial exhibition. Their book *Progettare scuole tra pedagogia e architettura*[15] *(Planning schools between pedagogy and architecture)* aims at an interdisciplinary language that could be a reference point for those wishing to construct or transform schools, convinced that if pedagogy lost its features as an abstract (and often inaccessible) subject it could move closer to the reality of school and school life via a relationship with architecture. *"At this moment* – Attia and Weyland state – *the strength pedagogy still lacks to address the school may find some answers in architecture, which reasons all-round on the construction of its buildings"*: developing pedagogies does not mean thinking of didactics as a simple set of contents, but demands a vaster, more coherent reflection on the architecture accommodating them. Still today the relationship between educational studies and architecture enjoys little consideration, in spite of the clear suggestions of great educators and great architects, who have highlighted the influence of space on learning processes. As Weyland points out, space design for learning needs the educational and architectural projects to interweave, to develop through a common vocabulary (to be shared with other disciplines, such as philosophical and cultural anthropology, aesthetics and psychology)

---

14. *Di ogni ordine e grado. L'architecttura della Scuola,* exhibition curated by Massimo Ferrari and Claudia Tinazzi with Beatrice Gerli and Daniela Maurer, ideated and produced by the Milan Triennial, Lazzaretto di Cagliari, 8-31 July 2016, local organisation by SPIN-APP Spazi innovativi per l'apprendimento, Scientific Coordinator Massim Faiferri.
15. Sandy Attia and Beate Weyland, *Progettare scuole tra pedagogia e architettura,* Guerini, Milan 2015.

and be able to turn into a pedagogical instrument in the hands of teachers, helping them in their relationship with students, as they will know how to speak to their emotions. For his part, Lucien Kroll explains how space planning is strictly connected with the idea of inhabiting: *"School spaces need to become poetic, emotional, sentimental, to establish a vital relationship with the students inhabiting them, people growing and bringing their own sensibility [...] It is possible to support education if we are capable of 'poeticising' Spaces, giving citizenship to subjective living, recognising that inhabiting a place – therefore also a school – is an emotion that cannot be reduced to set of functions, or fixed as though it were a pattern of quantifiable requirements"*[16]. Inhabiting, that is, does not solely mean to occupy a space – it means to appropriate it, feeling at home; only by inhabiting a space is it possible to develop it: the active contribution of the people inhabiting a school space is decisive in activating its transformation process.

Spaces and shapes of educational communities

"The translation of the concepts public *and* private *in terms of differentiated responsibilities makes it easier for the architect to establish in which areas elements need to be put in place that will allow the users/inhabitants to contribute to the environmental project and in which areas this is less important. In organising a floor plan, at the moment the plan and section are designed and also on the basis of the* principle for installations, *the conditions may be created for a greater sense of responsibility and consequently also for greater involvement in the organisation and furnishing of an area. Only then will the users become inhabitants"*[17]. The chapter *Da utente ad abitante (From user to inhabitant)* of *Lessons for Students in Architecture,* Herman Hertzberger's first book, begins thus. The schools he has designed, mostly for Montessori institutes, reveal a depth of thought that supersedes the themes linked with education, to define the universal key terms all architecture has to deal with as design for the space of man's life in a democratic society. Not by chance was the first of the buildings chosen by Hertzberger to illustrate this theme the Montessori School he designed in Delft (1960-66), where he tried to give shape and life to the metaphor of an urban landscape composed of a series of 'little houses' (classrooms), placed along a common street (the hall). The teacher (the 'mother' for each single 'house') was entrusted with the task of defining the domestic shape and atmosphere of the interior together with the children. On this occasion Hertzberger reappraised the idea of classroom, transforming it to respond to the requirements of the Montessori method more than had been done up to then. The school spaces at Delft try to express educational principles based on fulfilling domestic duties, with the objective of strengthening the children's emotional affinity with the context; for the child, keeping his own 'house' (classroom) tidy means establishing an affective link with the environment he has to attend (school). The spatial structure deployed by Hertzberger enabled his 'inhabitants' (pupils) to 'feel at home', with elements of everyday home life (e.g. plants in suitable niches) or exhibits of their manual work (displayed in showcases), making the 'school-home' environments (classrooms) homely. This form of appropriation allows pupils to be an active part in the construction of environments in which they identify themselves as members of a community.

---

[16]. Lucien Kroll, *Tutto è paesaggio,* "Universale di Architettura", Testo e Immagine, Turin 1999, p. 58.
[17]. Herman Hertzberger, *Lessons for Students in Architecture,* 010, Rotterdam 1991, Italian trans. *Lezioni di Architettura,* Laterza, Bari 1996, p. 22.

In other school projects, too, Hertzberger developed the same design elements, able to cope with contemporary problems, by configuring flexible environments in relation to the uses and features the individual user (pupil) wished to organise within them. This approach could be compared with the concept of 'open work': the outcome, mostly, of a specific idea of architecture and the character of the work; Umberto Eco clarified the concept of 'open work' referring to the artifice of 'open modules', able to accommodate mutations: 'open work' means 'possibility' and therefore the setting up of a field of interpretative virtualities that are basically indefinite and available for various relations[18].
Pursuing this objective, architecture manages, on one hand, to represent the space of experience – active in the interests of the children and necessary for learning linked with the complex sequence of movements and relations in time and space to carry out a task – and, on the other, favours the process of identification/appropriation by inhabitants. *"The more influence you can personally have over the things that surround you, the more emotionally involved you will feel and the more attention you will give them; you will also be more ready to lavish love and care on the things around you. Affection can only be developed for those things with which one can identify oneself, things upon which one can project so much of one's identity and invest so much care and devotion that they become a part of oneself"*[19]. In other words, the child manages to interpret his role of 'inhabitant' of the school building if he feels involved in defining the spaces and shapes of his 'little house', with continuous cross-referencing between the parties, based on the reciprocity of roles: user and form are reinforced and interact. *"This kind of relation is similar to that between individual and community. Users launch themselves into the shape - just as individuals show their true colours in their variety of relations with others - while interacting, and therefore become what they are"*. Architecture, too, particularly through its shapes, can really act as an interpreter of the legitimate educational needs of a developing individual: spaces that turn into places leading to the endless possible situations of collective interaction, instead of spaces thought up solely for the acquisition of notions – gusts of life, anticipating what the student will face outside school. *"At times school is simply seen as the instrument for passing on a certain maximum quantity of knowledge to the generation that is developing. But this is not so. Knowledge is something that is dead; whereas school helps us to live"*[20]. The spaces of school buildings need to be imagined as 'places' where the individual personalities of learners can recognise themselves as active parts of a community: changing spaces, anticipating (if possible) urbanity that is as complex as it is open to social aggregation phenomena, rather than environments aimed at carrying out *"a central role in the progressive process of deterioration in ways of civil cohabiting and, even more, in fostering the harsh way mechanisms of social exclusion mechanisms"*[21], as also happens with our scholastic system. 'Inhabiting school' should mean learning to 'be in the world', to relate to others, obey rules, taking care of the space we are immersed in and belong to. The spatial component of education reveals an ethical foundation that, in the light of contemporary social behaviour in our country, struggles to find designers able to propose suitable, coherent models.

---

[18]. Umberto Eco, *Opera aperta, Forma e indeterminazione nelle poetiche contemporanee,* Bompiani, Milan 1962.
[19]. Herman Hertzberger, *Identity,* in "Forum", 7, 1967.
[20]. Albert Einstein, *Out of my later years,* Philosophical Library, New York 1950, Italian trans. *Pensieri degli anni difficili,* Boringhieri, Turin 1963, p. 79.
[21]. Francesco Dal Co, *Andare a scuola,* in "Casabella", 750-751, December 2006 – January 2007, p. 3.

Recent works by Marco Iacoboni[22], known for his development of research on Rizzolatti's mirror neurons, tell of the opposition between implicit and explicit thoughts, maintaining that biology provides us with empathy towards others; empathy and its consequences have filtered through our cultural types and specificities, causing reactions and perceptions that are different from simple sensory stimuli: cultural-emotional affinities favour and facilitate the process of empathy and optimise its efficacy. These results are a further confirmation of the essential, irreplaceable function of the school as a 'common place', within which different generations, as well as various kinds of living and thinking and cultures, can encounter each other and converse.

Research and design experiments

Although this potential of school spaces is used too little, recent experiments have constituted possible reference models for designing school buildings in Italy and the world. Some of the architects who were part of the group of lecturers at the 2016 ILS Summer School have been working for years on these themes; this book has brought together some of their creations, proposing them as a possible reference base for designing learning spaces.
As has been seen, social changes, technological innovation and rapid global access to information seem to have consolidated an outlook that has not found answers in the confined environments of traditional school buildings, but in fact claims new spaces for education, capable of generating a closer bond with the external world. Design for educational space has to tackle a series of themes that permeate the different meanings of teaching and learning in the contemporary society, first and foremost the reciprocity of learning and teaching, as elements connected with the urban dimension of the city. A new challenge is taking shape for school building planning: the relations between internal and external, the intermediate spaces and different spatial components are structured on models based on the creation of complex interactions, which will enable the building to function as a community place and facility for education on the territory.
Some of Patrick Fransen's projects seem to pursue these principles. His new building for the San Giuseppe Institute in Geel is indeed organised around a clear, simple urban strategy: to outline an open/covered space within the campus. This concept, in a nutshell, was developed in a project that raised the building, placing the classrooms on the upper floor, to maintain the public nature of the underlying space. The Waterrijk elementary school project is also part of a larger urban redevelopment plan, in which the San Giuseppe Institute principle is overturned: the main part of the school is placed under ground level, leaving part of the overlying space accessible also to local children. Finally, the new school building for Savio-Hilmare school is a hybrid between old and new. The gym, a part of the pre-existing building, has been turned into a multi-functional hall, while the new construction features a large roof, towering over two floors with a transparent façade. The result is a bright, airy school, seeking a relationship between inside and outside, filtered through the vast glass surfaces. In these three projects by Fransen, the school building is no longer a framework with rigid, functional spatial devices, characterised by closed floors and long corridors, but is conceived as a neutral structure, with the maximum flexibility possible, able to respond to the pedagogical evolution of the school and strategically integrated into the urban fibre. The school thus becomes a place of everyday life,

[22]. Marco Iacoboni, *I neuroni specchio. Come capiamo ciò che fanno gli altri*, Bollati Boringhieri, Turin 2008.

an urban scheme able to accommodate the hopes, wishes and requirements of the community it is helping to build.
Open space, especially green space projects, is one of the main themes in Gianluigi Mondaini's approach: used in both internal and external variants, it is able to generate clever oppositions of solidity and lightness, enhanced by transparency, with contemporary views of artificiality/naturality. The green may become authentic design material, both from the point of view of pedagogical action and as an efficient power-saving strategy. In his project for the *G. Magagnini* primary school at Montecarotto, but also for that of Staranzano crèche, it is the colour that accompanies the free forms and spatial sequences of the interiors, generating positive strategies to enter into empathic relations with the young users. In these projects colour and graphics, texts and images become an opportunity for mediation between space and user, namely resources able to reduce the distances with the place, making it welcoming and familiar.
Another of the 2016 ILS Summer School lecturers' projects tried to reconstruct a dialectical relationship with the place where it was located: the scholastic complex designed by the 5+1 AA in the municipality of Zugliano which they conceived around the idea that *"planning, building a school is a public event"* and that a school building may also and above all be a meeting place for people. The isolated building of the elementary school makes its presence felt by its roof, which rises, becomes coloured then breaks, marking out an internal courtyard and so doing *"is built into the place and builds the place"*. Around this cosy, protected space the classrooms, service spaces and large public gym are spread: *"The ground plan expresses an educational-pedagogical idea wherein the central space brings together the different educational requirements, sharing of experiences, group-work, the need to "be a community" [...] The courtyard becomes a symbol, in the etymological meaning of 'bring together', inside and outside, school and territory"*[23].
Sandy Attia and Matteo Scagnol's school complex made up of primary school, public library with infants' centre opposite, designed in Bolzano, is another interesting case-study on the theme of learning spaces. The dialectic set up by the comparison between the stereometric, right-angled shapes of the primary school and the curved ones of the nursery school complex characterise this urban piece, marked by the morphological contrast between the highly structured buildings of the district and the natural condition of the public park shapes and countryside around. The school complex consists of two buildings close to each other, which create a necessary dialectical condition, based on the dialogue between opposites; the design of the piazza favours a dialogue with the design of regular stone material, in which green islands stand out and trees projecting shade and intimacy, in the wide spread of public space. Also in this case the school complex takes on an urban role, establishing itself as an authentic catalyst for the social and cultural organisation of the city. The request for its presence and recognisability in the space of the city is the symptom of a deep, imperative need of society: the school becomes a *"fully-fledged place of culture, a place where those who want to know, learn, seek, understand are always welcome. With consequent new attention for the presence, namely for the look of a building which, be it new or renovated, communicates its meaning to the outside"*[24].

---

[23]. Samanta Bartocci and Fabrizio Pusceddu, *Come in una radura. Zugliano nuovo complesso scolastico,* in "Area", 146, May-June 2016, p. 113.
[24]. Sandy Attia and Beate Weyland, *Il corpo della scuola come punto di contatto del rapporto tra pedagogia e architettura,* in "FAMagazine", July-September 2016.

In recent years schools have been designed and realised in different parts of the world that offer a panorama of spatial solutions able to pursue the objective not just of a more exciting, communicative school, immersed in urban reality, but above all a school where there is coherence between pedagogical and spatial aspects. Various experiments, especially some constructions in developing countries, may enrich the field of school building with new ideas and pointers. A particularly significant case is the Handmade School in Rudrapur, Bangladesh, designed by Anna Heringer and Eike Roswag. Situated in the north of one of the most densely-populated regions of the country (and the world), it was built following traditional construction techniques, with the added know-how of collaborators with a German background, which provided the opportunity to train local labour. The Handmade School was the outcome of cooperation between the Meti (Modern Educational Training Institute), Dipshikka (a local non-governmental organisation) and two German auxiliary organisations. In a manner assimilating the Montessori pedagogical concepts, the lay-out of the ground floor plan favours both self-learning and group teaching, based on the principle of transmission of knowledge and abilities. Behind facades 50 cm thick (made of dried mud) three classrooms for group lessons rise on foundations of layered straw and mud; the elementary right-angled nature of the plan stretches into the organic winding of cavernous recesses, both in the classrooms for free work and in those for groupwork and work needing concentration. The trade-mark features of the school evade mere issues of construction technology, focusing on specifically spatial components that are able to facilitate the educational methods chosen. The contrast between the geometry of the classrooms (collective learning environments) and the elegant organic structure of the 'homes' for learners emerges in the plans. Continuous reference to the exterior of the building is clear, sometimes in direct connection with the interior, and sometimes filtered through elements able to visually control the relationships (coloured partitions on the ground floor, bamboo canopies on the first floor), accompanied by construction elements meant for something different on the part of users (thresholds and steps that become seats, for example). On the whole, the school constitutes a possible contemporary synthesis between universal design problems and local requirements. Participatory self-construction is enriched with Western technological knowledge, while the local community is the protagonist of the construction of a public building, as the school is, upon which their expectations are lavished. *"For us architecture has a particularly strong tie with the development of identity and self-awareness.* - Anne Heringen made clear - *The only imported resource, apart from donations, was technological knowledge. The next step for the inhabitants of Rudrapur will be to put the knowledge acquired into practice".* What public building can express these values if not (and at best) a school?

Energy, materials and new information technologies

There is undoubtedly a notable basic misunderstanding in the relationship between the school building and new information technologies. If, on the one hand, there is widespread awareness of the importance of these technologies in our society and we share the need to introduce the opportunities offered by them into teaching processes, on the other, it has not really been understood how to fully exploit these privileges. For the idea prevails to resort to the advantages offered to teaching processes by these innovations, without proper understanding of the need to reappraise the organisational and spatial models of the learning environment. The most significant case is represented by LIMs, introduced and often used in classes as an updated version of the old blackboard. The same principle applies to the use of computers, often allotted to spaces external to that where normal teaching takes place, to laboratories or special classrooms specifically dedicated to 'machines' and their world of rules, alien to the process of socialisation and organisation necessary to acquire skills. This process is linked with the actual understanding of the impact of the new technologies on teaching processes; it can only develop through the absorption of the paradigm constituted by the network concept, i.e. distributed intelligence, which materialises in the Internet. Tradition individual learning could be enriched by collective and connective learning processes, with the enhancement of the capacity to discuss and communicate, and by mindful use of the opportunities offered by the world of the global network and its social media.

Clearly, to be able to achieve these objectives, it is essential that new ideas be imagined for spaces that are open and flexible, no longer tied to the traditional class lesson but with a framework that will foster and develop learning linked with socialisation, cooperation and collaboration between groups of learners and learners/teachers: spaces amid real and virtual, suitable to encourage the exchange of information, communication and knowledge. A physical environment must be created that is capable of hosting an unlimited quantity of virtual environments able to enhance it and put it online, widening the concept of space for learning. Spaces and places of learning will no longer be delimited by the right-angled walls of traditional school buildings, but will open up and extend, integrating artificial and virtual reality in the physical one of a school that wants to give its pupils the ability to master with critical sense the wide range of cognitive potential they have available.

To extend the space of the school building does not mean that real space does not need to be structured however with materials and technologies fit to make it pleasant and comfortable. Studies aimed at implementing and developing ICT technologies are moving in this direction, with the dual purpose of improving the internal environment (in terms of microclimate conditions, brightness, temperature, functionality and safety, etc.) and providing devices and applications with didactic-educational goals. Faced with the obsolescence of Italian schools, these aspects should have a key function for steering new construction of school complexes, as should technical/energy/safety adaptation of the buildings comprising the vast national patrimony. The contributions collected in the Focus Groups organised during the 2016 ILS Summer School[25] point out the importance of energy innovation and safety in planning school buildings. The energy requirement of these buildings must be evaluated as the sum of the energy contributions relating to needs for heating, cooling, domestic hot water production, ventilation, lighting and transportation of people. This involves greater attention to the relationship between choices that concern the 'wrapping' and those linked with space type – choices of plant, in particular, must be made bearing in mind the local climate. Recent studies have shown the importance of the use of passive energy saving techniques, with special attention to the possible synergy between materials and construction technologies of the 'wrapping' and the space-volume configuration, able to considerably reduce energy consumption both in new builds and buildings for renovation. These techniques may be coupled with the use both of systems like green roofs, so as to reduce energy consumption during the summer months, and recycled materials that ensure durability over time.

---

25. 2016 ILS Summer School, in particular the Focus Group *Energia, materiali e nuove tecnologie* by Giancarlo Della Ventura, Giovanni Mazzitelli, Luca Malfatti and Antonello Monsù Scolaro, coordinated by Augusto Marcelli.

To recover and adapt the existing building patrimony, above all in Italy, a further objective is provided, linked with the vulnerability typical of the static-constructor features of buildings now outdated, which requires innovative proposals for seismic retrofitting able to define technological innovation elements consistent with the architectural and energy aspects of the building. The ultimate aim is to find a modular structural scheme that is at the same time seismically effective, to be part of the architectural and plant solutions without being an element of contrast but of integration in the planning of learning space.

Internal spatiality conceived and organised, thanks also and above all to the functional and perceptual qualities of the materials, becomes a key element for defining the quality of school environments in terms of aesthetics and wellbeing. These aspects are the subject of much discussion in modern pedagogical thought, attempting to identify categories able to define the qualities of school life. Planning comfort in school environments is a pedagogical necessity; it corresponds to the natural need to feel well, and improves the standards of life of those who spend a large part of the day in these buildings. Like the external elevations the interior of a school building also communicates. It is able to recount and echo, by means of the different spatial configurations and variations in its surfaces. In planning learning space, solutions should take into account the acoustic qualities of the different environments: depending on needs and activities, and also on the number present, communication may take on a higher or lower tone. The configuration of the environments and use of materials should therefore be combined in relation to the modulation of the acoustic reverberation, envisaging a variety of sonorous experiences resulting from the activities.

The school for each and every one

To reconsider the configuration of learning spaces, freeing them from the traditional models of organisation of didactics (stagnant in their respect for outdated parameters), and to develop this configuration with new methodologies (based on the convergence of themes, aspects and contents of different disciplines), also means to pay specific attention to the needs of students with various problems (intellectual disabilities, motor problems, sensory disorders, etc.). It is essential to define and structure spaces with a new conception, open and flexible, suitable to encourage and enable effective cooperation for and between the pupils. Planning has to give the right answers as regards the rights of people with disabilities, promoting maximum individual autonomy and aiming at integration into the common classes for everyone and for every cycle, as envisaged by Law 104 of 1992. In the 2016 ILS Summer School a Focus Group[26] worked on these themes to identify the contributions architecture might provide; Donatella Petretto's contribution was significant (presented, with the help of other scholars, in the essay published in this book[27]). It highlighted the studies that for some time have concentrated on strategies for inclusive environments, developed initially with the tendency to make 'special' changes to common environments. The Director of the Center for Accessible Housing of North Carolina State University, the architect and product designer Ronald Mace, suggested the

---

26. 2016 ILS Summer School Focus Group, *La scuola per tutti e per ognuno* by Paolo Garau, G. Filippo Dettori and Donatella Petretto, coordinated by Alessandro Zuddas.

27. Donatella Rita Petretto, Paola Piras, Ilenia Pistis, Carmelo Masala, *Disabilità, difficoltà di apprendimento ed ambiente. La psicologia della disabilità e la neuropsicologia quali basi per lo sviluppo di una "School for all"?*

notion of Universal Design as "planning of products and environments able to be used by everybody, to the greatest extent possible, with no need for adaptation or special aids".
This concept was developed in the field of learning by David Rose and Anne Meyer of the CAST Center for Applied Special Technology of Wakefield, taking the name of Universal Design for Learning, and aiming at inclusive environments for everyone. The need arises from these studies to explore further the *"relationship between the influence a neuropsychological reading (based mainly on the analysis of neuropsychological functions and cognitive processes) could have on the reflections on the (social and physical) role of the learning environment in revealing these difficulties or, vice versa, on enabling the development of the potential also of students with learning difficulties"*. Deeper knowledge may develop only with a profound change in paradigm: shifting the attention in general from a person's accountability to the study of a person's functioning - in the field of learning space design, from the specific neurological features of the student to their interaction with the learning environment. *"Neither does the passage from introduction, integration and then inclusion have just a terminological value but, as usual, semantics involves a model of approach and an anthropological perspective and has endorsed the importance of interaction between the person and the environment and between the person and people (who learn and teach)"*. This approach could give answers, for example, to the cases quoted by Alessandro Zuddas, who, as a member of our SPIN_APP[28] research group, investigated the methods of interaction with space and people (peers and adults concerned) of subjects with two types of disorder: autism spectrum disorders (ASD) and attention deficit hyperactivity disorder (ADHD). The direct correlation emerged between ASD and states of tension linked with unfamiliar, difficult to predict environments. The deficits of mentalisation and internal coherence in children with ASD, apart from making communication complicated, can also explain difficulties in perceiving time and space. In these cases the predictable nature of the environment, together with the rapid but modulated presentation of stimuli, becomes essential. Zuddas suggests some solutions, such as class-spaces with rooms adjacent to the main classroom, differentiated so as to permit temporary methods of personalised teaching that will help the child in difficulty to determine, with safety and predictability, the activities to carry out. Also in these cases attention both to chromaticism and other sensory stimuli like noises, and to the brightness of the rooms and gradation of light is essential.
The case-study presented by Josep Mias during the 2016 ILS Summer School seems to really pursue these objectives. The psychiatric unit of Arenys, Can Zariquiey, is the project for a place of treatment for young people with mental illness, the main therapy revolving around the cultivation and care of orchids, an activity to be carried out with great precision, able to considerably improve the patients' psychiatric condition. The building was thought up as a small, cosy school, where the young people were educated as specialists in orchid cultivation; inside, a sales point was set up for the plants, which gave further value to the work of the patients. The overall image of the centre recalls the specific relationship between space, learning and disability, demonstrating how it is possible to find a successful amalgamation in a pleasant, functional building.

---

[28]. SPIN_APP Spazi Innovativi per l'Apprendimento is research funded by the Sardinian Region under Regional Law 7, *Promozione della ricerca scientifica e dell'innovazione tecnologica in Sardegna*, Scientific Supervisor Massimo Faiferri.

One of the research objectives worked on during the 2016 ILS Summer School was to understand how much these special functions can be influenced by the different features of environments and how it is possible to design inclusive learning environments based on a complex interactive model, able to take into account the different learning methods connected with difficulties or disabilities of various kinds.

Innovative spaces for learning

Although pedagogy has recognised the artificiality of the separation of mind and body, word and movement, cognition and emotion, world of life and world of education, the organisation of training spaces, if not elsewhere in Italy, has for a long time been basically conceived for speaking to students' heads, excluding their bodies, emotions and feelings.

Silvano Tagliagambe[29] has pointed out the need for didactic planning as an open operation, available for the activation of multiple, interactive pathways, enriched with moments of individual and collective reflection. Tagliagambe emphasises, in particular, the opportunity provided by constructionist theories in psychology, which propose learning as reconstruction, rather than transmission of knowledge, and then introduces the concept of cognitive artefacts, namely objects and devices that facilitate the development of specific learning offered to us by constructionist theories. Very briefly, the relationship between cognitive processes and cognitive artefacts is a fundamental issue, inescapable for handling the question of the nature and efficacy of teaching processes. *"The basic idea is that it should be the learning environment itself that is made available, and the system itself of materials offered and teaching activities promoted, that will trigger an investigative process important for the learner, whose experience must be based on a process of continuous, flexible restructuring of pre-existing knowledge in accordance with needs that arise, from time to time, from new educational situations"*. The main reference seems to be to Jacques Derrida's concept of deconstruction, in the awareness that one cannot "deconstruct" without thinking of alternative reconstruction that will propose the introduction of what was the outcome of the deconstruction operation, in a different relational fibre. The student must be given greater responsibility, involving him in the reconstruction process of knowledge, sharing objectives - collective and individual - and attention to his particular inclinations and cognitive styles. Tagliagambe proposes *"the perspective be superseded of teaching processes characterised by the deconstruction of their traditional disciplinary base and subsequent 'reconstruction' of the relational fibre between notions and concepts, complying with objectives selected on the grounds of the didactic plan to be implemented"*. This is a perceptual-motor method of learning that, in contrast with the symbolic-reconstructive way of learning, does not take place just with the interpretation of texts or mere mental reconstruction, but through perception and motor action on reality. In this sense spatiality may be imagined as meant not solely as a container of something but, rather, coinciding with the concept of 'field', in the sense Berger attributes to the word, namely 'space awaiting events' and, at the same time, 'event in itself'. The existence of the field is the pre-condition for testing the learning that will take place when it is possible to prefigure the action projects that can be undertaken between us, our body and the world surrounding us, experimenting with complex interaction strategies. An innovative space for learning should be organised for action,

---

29. Silvano Tagliagambe, *Idea di scuola*, Antonio Tombolini Editore, Loreto 2016.

becoming a space designed by the architect, but also by the teacher, and developed by the learner: a system of free spaces, active, stimulating environments, full of things to discover and equipped to accommodate gestures and actions, open to various uses and fruition, thus to a wide range of possibilities that will make the experience of relations between students and activities with things, rich, intense and significant. The studies and experiments of the ILS research group move within the reflections summarised in this text, seeking spatiality able to act as a linking element between pedagogy and architecture, on common working ground, the ultimate aim of which is the setting up of an environment that will allow the child to express and experiment with his own constructive strength. This place should be a synthesis between the teacher's words and the architect's design to be able to cope with the different viewpoints (social, pedagogical, economic, technical, aesthetic) underpinning Rogers' educating architecture: *"discipline and freedom are dialectical terms amid which the small scholastic society struggles to mature and introduce itself into the vaster human consortium: in a few cases, as in this theme of the school, architecture presents more obvious adhesion between form and content: the teacher's word and the architect's design collaborate equally in training the young person. The aesthetics of the building leave the imprint of a particular moral seal by means of a calculated psychological effect"*[30].

---

[30]. Op. cit.

1
Schools and new learning environments

In harmony with the educational methodology, school buildings have envisaged, both for junior and secondary schools, the planning of inflexible, functional spaces with closed floors and long corridors linking up a neat succession of rectangular classrooms. The classroom was the only place where knowledge was passed on. Nowadays, the School has become more and more an urban organism that cooperates in the construction of parts of the city, sometimes becoming itself a small articulate complex for public spaces completely open to the outside, or to cosier, almost domestic environments. We are witnessing a transition from the School understood as a total institution, with industrial-style educational methods, to the School as an everyday place, which meets the expectations, wishes and needs of the community it is helping to build. To intervene in designing scholastic space means to responsibly reappraise the role space actually covers in the dynamics of learning processes, as well as think about regenerating didactic organisation.

#school #everyday place #urban organism #city #inhabitant

Some school activities:
interaction spaces, classes, work tables and experiences in the territory

23

1.1

Fuzzy School. A philosopher's partisan glimpse into the future of the school

Fabio Bacchini*

Our schools will change dramatically in the next decades, independently of whether we design this change or not. What subjects can try to manage this process? A most obvious but superficial answer is that politics can. Metaphorically, politics can only affect the shape of the banks, not the chemical composition of the waters of the river flowing among them. The power to alter the latter is to be attributed to all the people working in the school: teachers, students, managers as well as architects. Yes, architects who design schools – whose activity of course cannot be reduced to designing the physical constructions where the educational activities take place.

This paper is nothing but a set of quick and badly argued notes aimed at applauding some of the directions that the school change might take. I will rely on a number of presuppositions:

(1) The freer the people active in a school, the better the school. I am convinced that freedom is healthy for the very functions a school must perform (though I will not engage here to try to specify what these functions are). I will offer here and there some clues of what my best arguments supporting this presupposition would be – but, if for no other reason than space limits, I will not justify it as it would deserve;

(2) Designing the future of the school is a daring common enterprise that teachers, students, managers and architects should collaborate in;

(3) We need a valiant, holistic and even radical design. Weak and cautious attempts will prove ruinous. This presupposition – not differently from the others – will not be adequately supported by arguments. I will assume it, simply. Let me just note that also arguments *against* it are noble and convincing: it would be sufficient to refer to Popper's idea that a piecemeal social engineering rather than broad holistic change is the gateway to good social reforms (Popper, 1944-45).

Moving around and laying down on the floor

In most Western educational systems, body movements are taken at best as neglectable, and at worst as blasphemous if made inside the school – unless they are performed in that expressly dedicated couple of hours per week named as 'physical education'. In the school of the future, a first keypoint must consist in attributing much more importance to body movement. This formulation is misleading, however, because one might think that I am just proposing to increase the number of hours per week for physical education (for evidence that physical activity and aerobic exercise physically remodel our

* **Fabio Bacchini, DADU, University of Sassari.**

brains for peak performance, see e.g. Keeley & Fox 2009; Ratey 2008; Ratey & Sattelmair 2012). What I am advocating is something different. Perhaps it is more correct to speak of increasing the teachers and students' freedom to move their bodies. In particular, it is completely wrong to build the very idea of how a human being should be educated on the premise that young members of our species should sit still for many hours on a wooden uncomfortable chair. Young humans – as any young animal – have a compelling instinct to move, jump and explore space. A school imposing stillness to their students is nothing but a prison – a repressive institution.

Of course there is a smart answer to my *j'accuse*: in order to learn efficiently one must focus his/her attention on what is to be learnt, and focusing one's attention is best done in absence of movement. Well, this is simply untrue. There is a huge evidence that not only the expression of one's thoughts, but also the very activities of thinking, reasoning and creating are improved by physical activity and even different kinds of body movements, from walking around in the bedroom to playing one's ear as if it were a guitar. We do think, discover, invent more and better if we can gesture, move an arm, stroll about or sip a tea (Cook et al. 2008; Oppezzo & Schwartz 2014; Slepian & Ambady 2012). Non-sedentary workspaces improve individual and group capacity to combine information to develop creative ideas and solve problems (Knight & Baer 2014). And of course we cognitively perform better when we are not uncomfortable or forced to maintain the same position for a long time. In consequence, students should be allowed to stand up from their chairs literally whenever they want in a well-designed school. They should be allowed to walk, bounce around, lay down on the floor, and sprawl out in a corner. Movement is a most important ally of our senses in exploring the world (e.g. Berthoz, 1997); the discovery of canonical neurons provides a revolutionary view of human perception where the perceived object is less shape, dimension and colour than it is a node of all the possible motor acts we may perform for interacting with the object itself (Rizzolatti & Sinigaglia, 2006); moving our body by using proprioceptive information is also means for producing the image of ourselves that we want to show to other people in order to express respect, friendship, love or rage, in a complex social dance (Montero, 2006). Definitely body movement is an important part of the process of our learning to deal with the world. The educational system should better acknowledge this fact rather than systematically neglecting it as a result of its paying homage to the mother of all harmful dichotomies, i.e. the Mind/Body opposition, and to one of her daughters, i.e. the Learning/Moving dichotomy itself. It is not just that physical activity improves children and adolescents' cognition and brain health (e.g. Hillman et al. 2014); it's that physical activity improves cognition if performed during the cognitive task (Oppezzo & Schwartz 2014; Slepian & Ambady 2012)!

A reasonable objection to what I am promoting is that, if the students are allowed to just stand up and go away (but where?) while the teacher is speaking, what we call 'school' vanishes away. Freedom to move seems to be releasable only at a very high cost. In fact nobody seems any longer capable to learn anything, since we can predict that each young student will be more strongly attracted by the many far-away mermaids than by the present lesson by his/her teacher. How can the school as we essentially conceive it survive at all if free movement of students is conceded? Lying on the floor might be perhaps tolerable. But walking and even running! How is it possible to incorporate all this into the school?

One side of the coin: getting rid of the teacher-centred lesson

I claimed that designing the school of the future is a common enterprise by teachers, students and architects. One fundamental action that requires the active participation of all these categories is that of eliminating (or better, "softening away") the teacher-centred lesson. Another is that of eliminating (or again, "softening away") the classroom. These are two sides of the same coin.

Let's start from the teacher-centred lesson. We need to promote a fresh pluralism concerning how it is that (young) people can learn. To do this, we need in turn to get away from the teacher-centred lesson where the teacher talks and the students listen, that is so typical of the Italian educational system. What characterises the 'teacher-centred lesson is the dogma that the focus of the students' attention must be on the teacher. This should be thought as just a particular case in a vast family of situations where various focuses of attention with different targets, intensity, goals, emotions, impermeability, duration and resilience are simultaneously distributed in space. Imagine that while the teacher talks to part of the students, other groups turn on further 'attentional events' gravitating towards a topic, a joke or a gag. This should be completely legal – and, above all, should be conceived as the fruitful result of safeguarded freedom rather than as *distraction* – the label I think it would at best deserve according to dominant ideology in most Western school systems. Some of these non-teacher-centred attentional events will have short life. Some of them, however, will persist and grow up. Some of them will merely be the effect of a mostly human necessity to change and do something else, while some others will be more complex and serve different functions. Many of them, I must admit, will depart from Kant in order to land on football actuality. In these cases I trust that the brilliant teacher will cultivate the art of entering into the heart of the new hurricane in order to masterfully absorbing back its vortex into the principal one. This is just one among the many capabilities that a good teacher must have which it is impossible to reduce to any set of instructions. (For example: "What are you talking about, you behind the coat rack? Football? Of course, Man United. There is no genius today playing in the team. Perhaps no genius has played in the team after George Best. Incidentally, do you think it fair that George Best got paid so much more than his teammates, provided that quite surely he did not work harder than any of them? Yes, he was more decisive than the others for winning a match. What would Rawls say here? What would Kant say?")

I agree that, if all this work must be done each time that someone starts talking of something else (in a situation where the student is even encouraged to do so!), teaching becomes perhaps the most difficult and hard profession in the world. But this is already so, since good teachers continuously act this way. This also implies that no student will ever get a time out. When I was a student, there was a vertical window of glass or semitransparent plastic next to the door of the classroom. When a student who had dared originate a non-teacher-centred attentional event got a time out, she had to spend ten or twenty minutes standing outside the classroom with her back leaning against the glass, so as to show that she was serving her sentence. (Note that also for what concerns punishment, spatial and social discourse always go together. In the Italian school system the climax of punishment is constituted by different degrees of social as well as spatial expulsion, from the time out from the classroom to a temporary or permanent removal from a group, the class or the school itself.)

In the school I am an advocate for, the vitality of both the insufferable pest and the undisciplined student are always positively exploited by the teachers and the classmates for a better learning process rather than squandered in a twenty-minutes boredom outside the door. This is a wise and

responsible form of recycling: an amount of energy is saved and converted into useful dynamics while it once was simply wasted.
Embracing attentional focuses' pluralism implies that everybody can personalise her day and decide what to do of her time. It is not forbidden to read Dickens (or, the Peanuts) during the math class, just as it is allowed to talk about the Grammy Awards during the lesson on the rise of ancient Rome. Indeed there are no longer *lessons* – rather each student will be focusing her attention either on the teacher or one of her classmates, depending on times and situations. Incidentally, another thing that a student should be allowed to personalise is her own space, i.e. a desk, a portion of the wall surface, a small corner in the building. But let us turn to the destiny of the classroom.

The other side of the coin: getting rid of the classroom

Since there will be moments when the attentional event is unique as well as moments when there are many attentional events of varied nature and magnitude, the architectural space must be thought of as capable of fitting all these different situations. It should be a jointed, well-structured, multilayered space, rich of prosceniums as well as of backstages, equally comfortable to a small crowd of twelve people having a chat, a group of three friends gossiping, a couple of confidants revealing secrets to each other, and a solitary person reading her novel (by the way, the school of the future should assign more importance to reading, and not just rhetorically – for example, by dedicating a full month to the silent individual reading of self-chosen novels – see Casati, 2013 for this kind of suggestion).
One important feature of the freedom I am advocating is that it should not bother anyone. In order for it to be so, we need locations that are still part of the larger space where students bring on their learning program, but at the same time are capable to offer privacy and physical distance from the "main stage". Those friends who are in need for a place where to cackle or download some of Picasso's paintings from the Internet do not want to leave that educational event taking place that morning at school: they just want to modify it according to their specific purposes – what requires some temporary autonomy also under the spatial and topological respect. I think that a fruitful category here is that of niche – not in a wall but in the main spaces of the school. Imagine smaller and larger niches of varied nature and shape crowding around the most obvious spatial components of the school such as the central "ecclesia". Each niche would allow and even nudge the students to display different activities with different kinds of people. (Incidentally, the concept of niche is a very interesting one also from the point of view of evolutionary biology; see e.g. Lewontin 1991).
Great attention should be put on the issue of acoustic isolation vs. permeability, which is central for determining the real nature of a niche. The very same niche may become, if acoustically isolated, a temporary "room on one's own", and, if acoustically permeable, a semiprivate stall at the theatre of the school. In any case, however, we get rid of the inside/outside Manicheism of the box-classroom, and find instead a fractal and varied space, with many holes in it, whose parts differ greatly with respect to their quality, affordances, privacy, undisturbability, joinability, controllability, capacity to offer visual or acoustic control of the outside, and so on. The niches – each different from the other – create the articulation of the architectural space needed to offer a truly ample range of possible actions (for an vivid illustration and a concrete realisation of this idea, see the Vittra Södermalm and Vittra Telefonplan Schools in Stockholm by the Rosan Bosch Studio). Moreover, from the social point of view the community is encouraged to try all possible group interactions, since ideally the space

should elicit so many collective responses that all possible combinations of students engaging in different temporary activities should be tried in the long period – all two-, three-, four-, five-person possible groups and so on. In other words, the architectural space is a device for actualising otherwise neglected social possibilities, which in turn will interact with the development of the individual personalities and enrich them. It is also to be remarked that such a complex, varied and multifaceted space is the perfect environment for multiple intelligences to flourish (Gardner 1983; 1999), since each distinct intelligence requires to be fostered and stimulated by different spatial affordances (Zene 2013). Just as the "all or nothing" character of the teacher-centred lesson blows up into a whole host of learning activities, so the "all or nothing" character of the classroom is dismantled into a multitude of in-part and in-some-degree spatial and psychological commitments and participations (see e.g. Rodriguez 2013). The same should happen to the school itself, via the creation of many intermediary zones whose status is being partly inside and partly outside the school. For instance there will be, on one hand, 80%-inside 20%-outside areas where the school members have free access to while external citizens can be admitted in at certain conditions – and, on the other hand, 30%-inside 70%-outside areas which will be part of the city, though not completely, since some scholastic activities can invade them at different times under the teachers' supervision. This allows the school to be truly open to the city, and vice versa.

Let us say a little more about the learning space. It is similar to a many-sided, protean, mysterious theatre that is not suitable for standard theatrical performances where once again we would be faced with the opposition among the one activity performed on the stage (the drama played by the actors, i.e. the teacher-centred lesson) and the one activity carried out in the parterre (the students/audience which passively listen and watch). On the contrary, this theatre is the ideal venue for the inconclusive rehearsals of a strange company of actors that sometimes are capable to converge on and act the same theatrical script (there will always be, however, someone who does not contribute and stays out of the picture), and more often do different things having little to do with one another – one actor performs a one-man show, another actor exercises her muscles, two of them discuss about Brecht's Epic Theatre, and so on. We must think of it as a zigzaging theatre with a rhizomatic backstage and movable sceneries, alternating dark and light, designed for rehearsals aimed at understanding Shakespeare and Ibsen rather than seriously performing their works. A theatre perfectly fitting this maverick, bibliophile company – supported by a few senior actor, though basically free – that a future class of students will have to consist in.

References

- Berthoz, Alain (1997), *Le Sens du Mouvement*, Odile Jacob, Paris; Eng. Transl. *The Brain's Sense of Movement*, Harvard University Press, Cambridge (Mass.), 2000.
- Casati, Roberto (2013), *Contro il Colonialismo Digitale. Istruzioni per Continuare a Leggere*, Laterza, Roma-Bari.
- Cook, Susan Wagner, Mitchell, Zachary & Goldin-Meadow, Susan (2008), "Gesturing Makes Learning Last", *Cognition* 106, 2: 1047-1058.
- Gardner, Howard (1983). *Frames of Mind: The Theory of Multiple Intelligences*, Basic Books, New York.
- Gardner, Howard (1999). *Intelligence Reframed: Multiple Intelligences for the 21st Century*, Basic Books, New York.
- Hillman, C.H. et al (2014), "Effects of the Fitkids Randomized Controlled Trial on Executive Control and Brain Function", *Pediatrics* 134, 4: e1063-e1071.
- Keeley, Thomas J.H. & Fox, Kenneth R. (2009), "The Impact of Physical Activity and Fitness on Academic

- Achievement and Cognitive Performance in Children", *International Review of Sport and Exercise Psychology* 2, 2: 198-214.
- Knight, Andrew P. & Baer, Markus (2014), "Get Up, Stand Up: The Effects of a Non-Sedentary Workspace on Information Elaboration and Group Performance", *Social Psychological and Personality Science* 5, 8: 910-917.
- Lewontin, Richard (1991), *Biology as Ideology: The Doctrine of DNA*, Harper Perennial, New York.
- Montero, Barbara (2006), *Proprioception As an Aesthetic Sense*, Journal of Aesthetics and Art Criticism, 64, pp. 231-242.
- Morsella, Ezequiel & Krauss, Robert M. (2004), "The Role of Gestures in Spatial Working Memory and Speech", *The American Journal of Psychology* 117, 3: 411-424.
- Oppezzo, Marily & Schwartz, Daniel L. (2014), "Give Your Ideas Some Legs: The Positive Effect of Walking on Creative Thinking", *Journal Of Experimental Psychology: Learning, Memory, And Cognition* 40, 4: 1142-1152.
- Popper, Karl R. (1944-45), *The Poverty of Historicism*, "Economica", 11 e 12; then Routledge and Kegan Paul, London, 1957.
- Ratey, John J. (2008), *Spark: The Revolutionary New Science of Exercise and the Brain*, Little, Brown and Company, New York.
- Ratey, John J. & Sattelmair, Jacob (2012), "The Mandate for Movement: Schools as Agents of Change", in Aleta L. Meyer and Thomas P. Gullotta (Eds.), *Physical Activity Across the Lifespan*, Springer, New York.
- Rizzolatti, Giacomo & Sinigaglia, Corrado (2006), *So Quel Che Fai. Il Cervello Che Agisce e i Neuroni Specchio*, Cortina, Milano; Eng. Transl. *Mirrors in the Brain. How Our Minds Share Actions and Emotions*, Oxford University Press, Oxford, 2008.
- Rodriguez, Sophia (2013) "«Can We Just Get Rid of the Classroom?» Thinking Space, Relationally", *Taboo* 13, 1: 97-111.
- Slepian, Michael L. & Ambady, Nalini (2012), "Fluid Movement and Creativity", *Journal of Experimental Psychology: General* 141, 4: 625-629.
- Zene, Andreina (2013), *Learnspace*, Master's Thesis, Department of Architecture Design and Urban Studies, University of Sassari (Italy).

Vittra Telefonplan School, Stockholm. Design: Rosan Bosch Studio, 2011. Photo: Kim Wendt.

1

2

3

4

5

1| Vittra Telefonplan School, Stockholm. Design: Rosan Bosch Studio, 2011. Photo: Kim Wendt.
2| Vittra Telefonplan School, Stockholm. Design: Rosan Bosch Studio, 2011. Photo: Kim Wendt.
3| Vittra Södermalm School, Stockholm. Design: Rosan Bosch Studio, 2012. Photo: Kim Wendt.
4| Vittra Södermalm School, Stockholm. Design: Rosan Bosch Studio, 2012. Photo: Kim Wendt.
5| Sheikh Zayed Private Academy, Abu Dhabi. Design: Rosan Bosch Studio, 2015. Photo: Kim Wendt.

1.2

Planning schools between pedagogy and architecture

Sandy Attia e Beate Weyland*

This essay provides an introduction to a research project conducted in Alto Adige on the relationship between pedagogy and architecture as part of the process that leads to the construction or renovation of a school building. The research project, "Between Pedagogy and Architecture", financed by the Free University of Bolzano's Faculty of Education, was carried out from 2010 to 2014 in collaboration with Sandy Attia and Matteo Scagnol, of ModusArchitects-Brixen (Weyland, Attia 2015).

There is a process at work in the world of education that seeks to renew educational spaces and didactics. The aim is to create a new culture of learning and pedagogical quality to school architecture, new interdisciplinary skills and well-being within educational facilities. Thinking about a school's "software" is beginning to be complemented by explorations on the "hardware" that shapes it (Iori 1999).

This research is based on the premise that the process of innovation in schools cannot be generalized, but, rather, is the result of very concrete ideas and actions in which both pedagogy and architecture play an equally essential role.

A school is a physical structure in which a community transmits and develops its cultural heritage. As such, it *is a text written by both the head teacher and teachers together with the architect*, aided by the local administration, which makes it possible. The factors that play a role in this writing process correspond to the ideas that have led to this research project.

Pedagogy and architecture come together in schools through action, with the urgency to build and the necessity to provide adequate pedagogic-didactic programs. And action implies choices, opinions, and agreements.

The architecture of pedagogy

This work aims to illustrate architecture's relationship with the broad field of pedagogy, understood as education science, that is, the analysis and reflection that has historically determined the critical practice of reason over reality in education (Scurati, 1997). Who, what, how and why we educate are the subject of pedagogy; a field in which schooling occupies a privileged position as it is the protagonist of modern day pedagogy and major approaches in contemporary educational thought typically focus on it (Ibid., 1997, p. 5).

The broad discipline of educational science comprises four areas: general pedagogy, which deals with developing concepts and proposals; history of pedagogy, which provides a basis for understanding the evolution of educational thought and current issues; didactics and special pedagogy, which gather, validate and test instruments and models for the practice of teaching; and experimental pedagogy, which

* Sandy Attia, Modusarchitects; Beate Weyland, Free University of Bozen-Bolzano.

concentrates on research methods. According to Paolo Calidoni (2000), the area of general pedagogy deals with the why and the needs of those who are to be formed by education; didactics with the how and what is possible. Experimental and methodological pedagogy deal with evaluating, on the basis of evidence, the effects of pedagogy's propositions, which didactics implements.
Although it mainly concerns the area of didactics, the relationship between pedagogy and architecture cannot neglect the other three, which prove fundamental in the planning process: Pedagogy offers the vision and a sense of the school's profile, providing a frame of reference made of principles, values, ideals and propositions. Pedagogy and didactics both aim to improve the development of education (Avanzini, 2006) and when they address the institutional setting where it takes place, they speak to and complement architecture.

The pedagogy of architecture

Architecture, including that for schools, plans and builds environments for human beings to live in. John Ruskin (1854) captured its aims and features in saying: "Architecture is the art of displaying and decorating buildings raised by humans for whatever purpose so that their presence contributes to the health, strength and fulfillment of the human spirit"(p.97). Architecture first and foremost provides an answer to the human primordial need at birth for protection and security, such as when Adam upon being cast out of the Garden of Eden and even before he used language, felt the need to cover himself and find shelter from the elements. In responding to the needs of every historical age, architecture is thus strongly influenced by the cultural, economic and social changes that are part of the human condition (Morris, 1881).
Architecture, like pedagogy, is a way of understanding and intervening upon reality. In this sense, it is subject to the ideas, research and actions that historically have dealt with the way spaces for education are to be conceived. If architecture is communication, that is, a written text, one that people (often unconsciously) read, discuss, criticize, and praise (Eco, 1968), then it also communicates the values interpreted by architects through experiencing, listening and understanding society, history, and human problems. The architecture of schools thus provides an interesting key for understanding modern-day discourse on the relationship between teaching and learning, as well as how thought on the sites of education have evolved.

The elements of a dual strategy

The combination of pedagogy and architecture provides added value to the planning of a school and opens interesting perspectives for the following reasons.
Firstly, it provides a more comprehensive understanding of social, historical and cultural complexity by focusing more closely on the role of memory and the past. School is traditionally the place where the keys to culture are passed on, guaranteeing access to everything human beings have understood, discovered, and invented in the past. In school we find the elements of our past, which is the foundation of what we intend to create in the future. In architecture, the reference to the past is a given, an unavoidable act: *"an act of memory projected in the future"* (Frediani, 2011, p. 21). The challenge in this sense is to build on the past with the aim of putting forward innovative, sustainable proposals for today.
Secondly, the pedagogy-architecture binomial allows us to reflect on the concrete, material aspect of schools. In architecture, to build means to use what is available, employing the manifold languages of

matter. Attention is especially focused on the quality of the materials, on their intrinsic meaning, on how they are perceived and experienced. In education, the relationship with words dealing with materials have, in contrast, almost an abstract value. To build pedagogies means to work with the raw material of knowledge; to practice didactics means to speak the languages of the various subjects of study and to be aware of their different qualities.
A third fundamental aspect of this intersection is the focus on the relationship between objects and people. The construction of a school generates a system of relationships in which the triad teacher-student-knowledge transforms into a framework composed of discourses and resonances that lead to the inclusion of space in the definition and organization of the educational relationship.
A further element that characterizes the relationship between pedagogy and architecture is the courage to choose. Architecture's task is to trace the boundaries between the various elements, to define limits and consequences, to divide space between interiors and exteriors, between the space you first encounter and that which follows. Didactics also works with knowledge by carefully selecting, analyzing and categorizing knowledge in specific learning units with corresponding learning materials in order to give form to knowledge. Each of these actions implies a choice and is determined by a specific planning pattern, both in architecture and didactics. Gianluca Frediani tells his students that, "Planning is a strenuous act of renouncing all that is superfluous. Renunciation is always painful. But this pain gives value to choice. To be means to exclude." (2011, p. 16). Today's task for schools is for each to choose its own pedagogic blueprint and be able to justify and support it in the educational and didactic process.

The neverending game

The relationship between pedagogy and architecture is challenging us on a number of issues: defining the relationship between space and didactics; finding solutions with a broad cultural perspective useful to the school's mission; meeting the challenge of illusion and of the continuous metamorphosis of perspectives on things and to aim at clear and well-defined choices that exclude all that "is not"; planning bright, multi-perspective, modern and sustainable environments, which respect the memory of things and people and are based on the silent and open interrogation of traces of the past, while imagining the future.
To plan a school, and not just build one, is never an easy challenge. It is a never-ending game, with an infinite succession of acts. It is a project for a time that will come. In this sense it is a constantly open and unfinished project, based on trust in the extraordinary potential of consciousness and responsibility as an expression of the efforts of head teachers, teachers, architects and authorities to create new cultural spaces for the community.

References

- Avanzini A. (2006). *Didactica*. Pisa: Ed. Del Cerro.
- Calidoni P. (2000). *Didattica come sapere professionale*. Brescia: La Scuola.
- Eco U. (1968). *La struttura assente*. Milano: Bompiani.
- Frediani G. (2011). *Architettura è*. Venezia: Cluva.
- Iori V. (1999). *Lo spazio vissuto*. Firenze: La Nuova Italia
- Morris W. (1985). *Prospects of Architecture in Civilisation* . In Manieri M. (Ed.). Opere. Bari: Laterza.
- Ruskin J. (2010). *Lectures on Architecture and Painting*. In Cook T.E, Wedderburn A.
- Scurati C. (1997). *Pedagogia della scuola*. Brescia: La Scuola.
- Weyland B., Attia S. (2015). *Progettare scuole tra pedagogia e architettura*. Milano: Guerini e Associati.

Designing schools, between pedagogy and architecture, teaching and design;
http://pro2.unibz.it/projects/blogs/pedagogiaearchitettura/

Experiences of research and project:

Meeting and symbolism: a space which educate

Valter Caldana*

> *School is, indeed, a place to learn. But learning implies the experience of life and, therefore, implies then all the activities of life, from work to leisure and often those of your own home." (Teixeira, 1964)*

Before planning the future we must know the past; we must know what we want and where we want to get, but above all, we must know how we can cruise past.
When we talking about to innovate spaces for learning, reflection arises: since the teaching/learning process has the primary function of preserving and transmitting knowledge, culture and ethical values, can the same process and its particular environment, be conservative by itself?
To pass on, we need a conservative space?
The cultural baggage that each society brings with it, dwells in places designated to transmit and disseminate knowledge. However, the idea of space for learning comes from an innate sense of solidarity among the peoples. Therefore, it is clear that in this view the learning spaces can not be conceived as places for a simple and notional conservation, as they were born primarily as a function of the public welfare.
Education is crucial in the organization of the social fabric and makes possible the inclusion of citizens in community issues. In addition to this, it's the strong educational investment which produces results on all scales and in all the variables that make up the complex organization of contemporary society. In the large barn of education, the seeds of health, work, leisure, home and security are preserved. Here they thrive, and strengthen the sense of belonging, solidarity and justice. This is where the creativity, skills and personal aptitude will be discovered, understood, encouraged and developed, removing in this way the children, young people and adults from *risk* situations.
When we talk about education, we are dealing, in a systemic way, all these aspects simultaneously; in terms of its complexity as a service, product, heritage and at the same time the intrinsic and inalienable value of citizenship construction process.
So, the teaching process belongs innately to public policy, which is what allows us to drive decisions and actions to get results. Its primary function is to build the future and strengthen the sense of citizenship in a democratic vision of knowledge. This means that knowledge must not only be accessible to all, but must also leave space to free conscience of citizens, in every moment of their lives. In this perspective, learning has a role whose purpose is to pose problems, which is to overcome

* Valter Caldana, architect, FAU Universidade Mackenzie, São Paolo, Brazil
The contribution was written in collaboration with AnnaMaria Laddomada.

the dogmatic and unique structure for which the teacher has the authority and the student is in a state of dependence.
Says Paulo Freire, a Brazilian educator, "the school is important not only to study, but also to create friendships and bonds of coexistence." (Freire, 1991)
For architects, learning spaces can not just be conceived as mere centers for the transmission of notions, but they must have the aim of training future aware citizens. Whatever the learning space is, the pedagogical political project is the main tool of organize architectural ideas and made interventions in the reality. *"School, in fact, includes economic investment among the highest [...], is one of the most complex architectural programs as it includes elements of human residence, food and health services, sports and recreation, spaces such as the library, museum, theater and auditorium, laboratories and warehouses, to say nothing of its activities and specific spaces, classrooms and laboratories. The school architecture, therefore, includes all the architectural genres."* (Teixeira, 1964)
Societies are constantly changing and that occidental in particular, in the last century, has created gaps so radical that still are incomplete and incômpressible. The existing school buildings are proof: we have classrooms of the nineteenth century, teachers of the twentieth and students of the XXI. An educational organization renewal is needed more than ever and must begin with a review of the dynamics of learning processes, which should require a greater awareness and active involvement of the students, a mutual relationship between students and teachers, between the two and the space and primarily between the three and the society and the contemporary world.
The learning space must therefore encourage the abilities of its users, who can no more be considered mere actors in a *deja vue* scene. We need to identify the potential of their conscience and the importance of their intentional relationship with the world, inside an active vision of knowledge.
This means that, for children, learning and its environment must establish an interactive relationship with reality, which can be served by the technology and especially through connectivity. But not only. If for the generations born up to the 80s and 90s there is a difference between technological interactivity and that physical experience, we must understand that for today's children and future generations does not happen the same. For the twenty-first century students there is no dichotomy between the *hi-tech* and human experience. One learns everything, all the time, by any means and in any place. In a pedagogical vision, learning spaces and technologies are at the same level.
For this reason, the educational environment and learning spaces, must be rethought in its own deep. Task of the architects is rethink them as no longer "useful space", functional, as an immediate response to a program of low or medium complexity, but even as "response space" to a set of educational directives, even if the most advanced possible, as for as the Freinet pedagogy, Steiner, Montessori, etc. At this time of discussion of values and implementation of limits, we find ourselves faced with an even swinging motion between development and relegation, in the balance between a strong rapprochement of the analog materiality to the digital immateriality. The contribution that architects can generate is to propose and draw spaces that can overcome these dichotomies. "The ideal project does not exist; in every project there is the opportunity to realize an approximation." (Rocha, 2016)
Perhaps it is time to understand the learning space, especially the scholastic space, as a meeting space, since it is precisely the encounter and the exchange that makes possible the relationships between people and that is what strengthens the civil and political dimension of this space, making it the "symbol space".

The key words for a change should be: city, space, learning, citizenship. So we need to understand that in the XXI century people live in the city and they learn with the city: the whole space becomes service and the public space becomes even public service. It is therefore necessary to recognize the symbolism of space and imagine new ones. "No one educates anyone, no one educates himself, men are educated each other mediated by the world." (Freire, 1991)

Imaginary school spaces

References

- Freire P. (1991), *A Educação na Cidade*, São Paulo, Cortez Editora.
- Rocha P. M. da, Entrevista. http://www.archdaily.com.br/br/783953/frases-paulo-mendes-da-rocha- e-o-proje-to-ideal
- Teixeira A. (1964), "Plano e Finanças da Educação", *Revista Brasileira de Estudos Pedagógicos*, v.41, n.93

New spaces for teaching

João Nunes*

We talk about teaching Architecture of spaces and how Architecture of spaces can help to improve the effectiveness of the teaching process.
The act of teaching starts from a relationship between people. A relationship based on the willingness to share an experience (a will that must be generous and disinterested) and, on the other hand, on the willingness to receive the gift of experiencing mistakes, setbacks and successes in an exchange that can allow progress in personal research processes, through the correction of errors at birth, the anticipation of failures, the saving of disappointments and anguish, and through the suggestion of shortcuts for long and painful paths.
Regardless of the intrinsic quality of the people involved in such a relationship, if the relationship fails, if you do not build what allows it to work smoothly and intensively in building bridges of communication that go where the word does not arrive, the teaching does not work. It is an essentially relational construction and it is the relationship that, by superimposing itself on the individual characteristics of the protagonists of the process from which the process itself results, can ensure the success or failure of the construction.
The idea of the predominance of the relationship over the object is an idea intrinsic to the very idea of Landscape.
When we talk about Landscape, we are talking about a collective and articulated construction, made of space and time, made of the continuous process of the physical inscription of signs corresponding to convictions always in constant mutation, year after year, generation after generation. We are talking about a concept linked to two different types of Architecture, to two different forms of relating to the world around us, through our ability to transform it.
The idea of transformation is always linked to a critical attitude. The whole process of transformation begins with a critical observation, with a dissatisfaction with the state of things, with a desire and will to improve something, to increase the efficiency of a process, its performance, the security of our group, family, tribe, country, comfort, or simply, the beauty of a concrete situation in which the world presents itself. The critical idea locates the will of transformation in a concrete cultural context, capable of organizing models of functioning, organization or image in an abstract way, i.e. as models that, by guiding criticism, orient transformation before its concrete physical formalization.
Of course, the models are altered in every moment according to the context, according to the paradigms of beauty, comfort, efficiency and safety possible in every cultural and historical moment, and the transformations are, therefore, determined as a consequence of each moment: everything that characterizes a moment and that animates it, everything that Men refer to in every moment of time is shaped in a more or less evident, more or less direct way, following what guides, and orients the transformations of that moment.

* João Nunes, Architettura del paesaggio PROAP.

This tectonic logic of the Landscape corresponds, on the one hand, to a genesis in which the successive transformations are configured in a continuous system of superimpositions, a stratification of signals that corresponds to the continuous succession of opinions on what orients the construction of models of beauty, comfort, efficiency, safety, according to an organization of successive superimpositions described by many sources with the metaphor of the palimpsest. On the other hand, the inconsistency of the process of continuous alteration of paradigms does not exclude partial regression, or the appreciation of previously denied values. The very character of the transformations, moreover, requires that a new layer cannot be recorded indifferently with respect to the lower layer, so that the construction of the Landscape is formulated through complex construction processes that are continuously reactive over time. It is therefore a Palimpsest, indeed, but in a way that guarantees that the superimposition of one layer on the other does not correspond to the cancellation of the first, but rather to a partial digestion, to a metamorphosis, at least partial. All this contributes to corroborate the observation that sees in the Landscape a construction made of both Time and Space.
Everything begins with our dissatisfaction with what the world offers us: initially a natural world to be transformed, then an anthropic world, in which the object of criticism will be rather the set of models that led the set of transformations of the previous generation. And so, animated by values different from those of our predecessors, by a different (possibly deeper) critical capacity, by scientific data that allow a vision of reality that seems more lucid (which will later be confirmed as erroneous), the will to transform orients the construction of a new layer of the world, a new set of signals destined to last as long as the integrity of values, convictions, the consistency of scientific knowledge that describes and interprets reality can last. Destined, therefore, to be replaced sooner or later by another layer.
The transformations that this process guides belong to two very different categories that, in turn, correspond to two very different ideas of Architecture.
On the one hand, it is an Architecture that starts from our body, from the measure of our body or our gestures, from what we individually do or in the groups in which we organize ourselves: the Architecture of a shell conceived in function of an individual or collective body and the spatial implications, the functions inherent in the functioning of that body. These processes are fundamentally linked to the idea of "manufacturing" in which we raise, according to the indications of our critical vision, the shells that we call houses, markets, cinemas, theatres, workshops, offices, factories and, of course, schools, or other places of education, with the intention of building a new world that we believe to be better.
And it is, on the other hand, an Architecture that has nothing to do with the measure of our body or with the measures related to our group functions and the corresponding spaces. An Architecture that starts from the measurement of the metabolisms of spontaneous functioning of things, of natural mechanisms, and that, through processes related to the idea of "domestication" proposes minimal and opportunistic transformations, able to exploit these functions to serve the objectives of comfort, safety, productivity of this new and better world that we want to see built.
Landscape is the meeting territory of these two Architectures that are characterized, as well, as complementary systems of manufacturing and domestication: they work with the rules of material manipulation in the production of new forms, but also putting their hand to the ability of conduction and orientation of the conditioned evolution of natural metabolisms.

The methods and procedures characteristic of each of these ways of understanding the transformation of the world, Landscape Architecture, are contrasting and allow to enunciate two methods different work that must coexist and integrate into the landscape project. If on the one hand the idea architectural presupposes the existence of a corresponding formal referent, in the traditional method of Architecture, to the drawing, or rather, to the drawn (what the drawing represents builds, is an objective of physical materialization that corresponds to a static model), on the other hand the concept of domestication juxtaposes this process with another process, evidently dynamic, evolutionary, intrinsically linked to time and the transformations that, in time, will be proposed, planned, or simply accepted. The admission of a physical change of the object of the transformation expressed, concretely, by the role that the volumes of the tree masses have in the definition of the architectural space in a landscape project (from almost non-existent at the beginning of the life of the project, until fundamental and dominant, when the project matures) adds to the idea the need to conceive, to design the spatial relations of a conventional Architecture of an objective architecture that continuously alter between everything that constitutes the project that is physically more stable (buildings, walls, stones, topography, water, ...) and what is changeable and that, changing, alters the configuration of the architectural space and its perception (shrubs, trees, deciduous and perennial leaves, light, transparency of the foliage, open and closed spaces, blooms and corresponding colours and scents, ...).

ETAR VALLEY OF ALCÂNTARA *

Etar water treatment station in Alcantara, Lisbon, Portugal 2011

Start: 01.2005
End: 01.2011
Infrastructure Integration: Environmental Redevelopment
Area: 8.4 ha
Client: Câmara Municipal de Lisboa
Authors: João Nunes, Carlos Ribas
Location: Lisbon, Portugal
Collaborators: Carla Silva, Clara Guedes, Joana Barreto, Mariana Sargo, Miguel Coelho de Sousa, Nuno Mota, Sara Neves
Architecture: Manuel Aires Mateus, Frederico Valsassina (FVA)

* The intervention develops territorial reconfiguration of the morphology of the Alcântara Valley and as a comparison between distinct and complementary characters - slope space / coverage space.
The design reconfiguration builds a path characterized by a landscape redevelopment of the Valley, in an attempt to recover a dialogue between the two constantly disconnected slopes and the re-establishment of a relationship built by the cut imposed by the current ETAR. We are talking about landscape redevelopment because the intent is to recover the continuities that have been lost, both from the point of view of the visual and ecological systems of the city landscape.

Administrative areas

Green roofing systems

PANORAMIC TERRACE OF SANTA CATARINA

Urban redevelopment of Alto de Santa Catarina, Lisbon, Portugal, 2013

Start: 2003
End: 2013
Urban Design, Parks and Gardens, Public Space
Area: 6015 mq
Client: Câmara Municipal de Lisboa
Location: Lisbon, Portugal
Collaborators: João Nunes, Carlos Ribas, Nuno Jacinto, Iñaki Zoilo, Sílvia Basílio, Miguel Sousa, Bernardo Faria

Playgrounds - Spaces for Play

Public Space

2
Energy, materials and new technologies

Although decades of research have observed the limits of the school model of industrial times, the so-called "cell & bells" model, it is clear that this type has by no means been surpassed; the one-way relation between teaching and learning remains dominant in educational practice. To move on from this model we need to start again from space, from a flexible, multi-purpose environment, where groups of students assemble, then split up depending on their interests and skills, strengthened by the introduction and use of the new ICT information and communication technologies that has to accompany a vertical learning process aimed at constructing and enhancing skills in a planned, continuing way. School space is the space of experience and must therefore be active towards learners.

#school #ICT #learning process #didactic organization #flexible space #modular and multifunctional #space of experience #child

Focus Group dedicated to energy, materials, new technologies.

2.1

Energy, materials and new technologies for innovative learning spaces

Augusto Marcelli*

In the world of education, e-learning is undoubtedly one of the topics with major potential for expansion and opportunity for benefit. However, we cannot yet imagine the replacement of schools made in brick and walls with virtual schools, open on line. The relationship with teachers and the availability of dedicated facilities for learning and stimulating the creativity remain essential components of the student training, not just of children. Still, reconsider the "spaces" dedicated to learning is a priority for the entire school system, and a challenge for "architecture" that should take into account both the great social changes underway and the new information and communication technologies available nowadays.

This contribution briefly summarizes a series of discussions within the Focus Group dedicated to Energy, Materials and New Technologies, and motivated since these concepts are quite often introduced and tackled improperly. The ongoing social progresses and the technological developments clearly highlight the limitations of the "school model" focused on an educational approach in which the relationship among teacher and students is confined within physical spaces, with no or limited space for creativity, and often preventing the use of new available technologies.

The school as building is a socially unchallenged structure that goes unnoticed in the urban system, hosting educational activities in *homely* spaces. If this was a winning model in the past, now it is no more representative and, above all not more adequate to expectations, desires and emerging demands of the society.

"It is the action that leads to the knowledge and the space clearly supports this process."[1] Today and in particular tomorrow school must truly be a "space", a ductile partner, open to the civil society and, ideally, with no barriers.

For the great possibilities and the foreseen economic advantages, *e-learning* and in general ICT (Information and Communication Technology) are areas in continuous expansion both in the Western countries and in the less industrialized ones. However, it will likely remain impossible for many decades to imagine a society with no physical relationships among teachers and students in dedicated spaces where learning and training are optimized. To this purpose, it is a must to reconsider these dedicated "spaces" and take into account the most advanced technological tools, with a priority to energy, environment and safety for the entire school system, which cannot continue ignoring social changes underway and extraordinary opportunities offered by ICT.

* **Augusto Marcelli, INFN, National Institute for Nuclear Physics.**

Information and Communication Technology (ICT)

ICT is the set of technologies that allow processing and communicating information through digital media such as computers, tablets and smartphones, nowadays tools available to both students and teachers. This smart technology allows acquisition and manipulation of data, including conversion, storage, protection, transmission and the secure recovery of large amounts of data. ICT is therefore an essential resource for institutions and organizations in need to manage large volumes of information quickly, effectively and efficiently. We may say that since 90s the impact of ICT on the rise of the productivity of our society can be linked to the shift of the technological paradigm associated to the most important revolutions in the history of modern economics as well as of the general-purpose technologies. ICT is a dynamic tool continuously evolving that in the recent years become more and more important, because it can generate further information and knowledge. It represents now an indispensable tool for the "modern" school, in particular if we consider new perspectives offered by merging technologies such as computers, telecommunications, electronics, and *media*. In the framework of educational space, these new opportunities will certainly lead to fundamental changes in the concept of "school" and, more generally, to the evolution of the concept of the "educational space".

Energy

In the present technological scenario, energy is an essential issue to consider for the design and the construction of the schools: buildings are distributed all along the country. However, in this context there are fundamental questions to answer.

Is nowadays it possible building "schools" that are energy independent and also eco-compatible? To reach this goal it is necessary to store energy available from all possible renewable sources: solar, hydro, geothermal and wind energy, and learn how to handle emergencies. Moreover, take advantage of the availability of the energy at low or no cost could make possible many educational and training activities, now inaccessible due to their costs.

The idea of energetically independent buildings requires modern and efficient energy storage systems, but also choice a different approach, e.g., using new materials, considering room lighting with a different perspective, introducing in the design efficient and reliable services as new opportunities. Reshape a country starting from schools may represent an extraordinary social and economic opportunity. A society that sets at its centre modern schools, energetically independent is certainly an advanced society that also takes care of the environment: a great place to educate younger generations to energy sustainability and environmental protection issues.

Environment

The environment is certainly one of the most challenging issues that a modern society must consider. The International Kyoto Protocol was born to limit the global warming phenomenon and certificated the role of anthropogenic emissions and its correlation to global greenhouse effect. By 2030, the Earth will account for about one billion people more, totalling circa 8.5 billion of persons mainly concentrated in large urban areas, a growth that will enhance pollution and shortage of many basic resources. The continuous increase of temperature and the dramatic air pollution not only in many urban areas will require in the future careful monitoring to make possible many outdoors recreational and sports activities. As a consequence, programs to monitor and reduce the particulate are becoming strategic:

in fact there are already available on the market systems to clean up the atmosphere by removing fine dust.[2] Modern technologies make now possible important advancements, and soon compact devices with a great sensitivity will be able to provide real time information on the air quality. A network of these compact mobile devices will be capable to offer a great deal of useful information.[3] However, pollution is also an indoor problem, at home, in working areas and in large public environments. The control of indoor air quality is already a must since many people spend almost 90% of their time indoors, and in many environments healthy problems have been recognized.
Despite the growing risk to health associated to the exposure of indoor pollutants, many countries have not yet considered this issue in their legislation. Despite several studies showed that considering offices, homes, schools, stations, shopping centers, etc., many of these indoor sites present negative effects on health, the air quality remains not regulated.[4] It is therefore necessary to educate people to environmental safety and to the pollution. Starting from schools a great effort can be done, in addition to research investments as already happens in the case of working environments.[5]

The school is an open space to the society, and for this reason it has and will continue to have for many years ahead a pivotal role in the growth of individuals and to their integration in the society. Designing "spaces" dedicated to learning is therefore not just a priority but an extraordinary interdisciplinary effort that involve the entire knowledge: from neurobiology to mathematics, from physics to engineering. The challenge to the "architecture" is how to consider and introduce in a reliable way all amazing ICT technologies now available. However, the capability to take advantage of these opportunities depends only on the possibilities of access to the net. This mechanism based on knowledge and skills in the use of technologies and communication networks will provide great social and economic returns.[6] In this regard it is important to underline that the advantage offered by ICT technologies and their effectiveness is a quadratic function of the number of users involved.[7]

---

1. F. Pusceddu (2010) *Spazio-Corpo-Mente: Lo spazio come luogo di invito all'azione*, Scuola di Dottorato in Architettura e Pianificazione XXIII Ciclo - Università degli Studi di Sassari
2. http://www.istechpot.com/progetti/progetto-apas-abbattimento-polveri-atmosferiche.html
3. F. Gozzi, G. Della Ventura, A. Marcelli, F. Lucci (2016) Pollution hazards from particulate matter (pm): state of the art and perspectives for pm monitoring, Atmospheric Pollution Research 7, 228-234
4. F. Gozzi, G. Della Ventura, A. Marcelli, F. Lucci (2017) Current Status of Particulate Matter Pollution in Europe and Future Perspectives: a Review, J. Mater. Environ. Sci. 8, 1901-1909
5. E. Anthes, The office experiment (2016) Nature 537, 294-296
6. A. Van der Meer, W. Van Winden (2003) Regional Studies E-governance in cities: a comparison of urban information and communication technology policies
7. M. Castells (1996) "The Information Age". Economy, Society and Culture. Vol. I: The Rise of the Network Society (Cornwall, Blackwell T.J. Ltd).

2.2

The school: a permanent and innovative laboratory for new Smart Technologies

Giancarlo Della Ventura*
e Alessandro Cecili

One of the central paradigms of the ILS school is stated in the introduction to the school booklet: "*The space in the school must be a space to develop experience, and for this purpose this space must to be active with respect to the students, already full of potential for knowledge*"; to accomplish this task, one needs to "*restart from a flexible and versatile environment where students can eventually group according to their interests and skills, and be supported by the use of new information and communications technology (ICT)*".

These sentences contain two basic concepts: the first concerns the use of new GIS/ICT technologies as a tool for the designer of new educational spaces; the second concerns the use of these technologies as innovative learning tools within these spaces.

The usefulness of the new technologies for the visual/virtual representation of the reality as a tool for the design is well consolidated. It is sufficient to consider the increasingly widespread use of, for example, laser scanning, which allows, in a relatively simple way, a 3D representation of the design environment and a virtual browsing within the spaces, besides a simple, interactive and quantitative evaluation of the volumes and working spaces. During the design phase, these data are of great utility for the evaluation, for example, of the expected costs both for the movement and for the implementation of the construction materials. The possibility to surf the high-resolution cloud of scanned points, possibly coupled with other data such as photogrammetry, optical scanner or infrared thermography, then allows very interesting applications for the documentation and maintenance of the spaces. It is now possible to directly transfer the 3D model, obtained via laser scanning, into a BIM (building information model), and manage the flow of data on AUTOCAD-type platforms, thanks to a complete compatibility between the information. In this way, geo-referenced models (as are those produced by modern scanners) can be accessed interactively and can be used as containers for all kinds of information for the designer. In addition, we can mention the possibility via GIS/webGIS, to construct high-resolution image banks for the historical and architectural documentation of entire sites, as well as of specific art works; these applications may become extremely important as a documentation for the original building in areas, like the Italian peninsula, characterized by a significant degrees of exposure to the geological risks (seismic, volcanic, or hydrogeological), or even the simple weathering. Several examples of the above applications can be found on the website of the GIS Laboratory of the University Roma Tre (http://host.uniroma3.it/laboratori/labgis/).

* Giancarlo Della Ventura and Alessandro Cecili, Laboratory GIS, Department of Science, University of Roma Tre.

ICT technologies may thus have a fundamental function within the innovative educational spaces, for at least two reasons: (i) for the active management of the space, and (ii) for experimentation and continuous learning, following the new IoT (Internet-of-Things) philosophy, according to which the student becomes the creator of the technologies that enable him to optimize his life.

(I) The new "Smart" technologies offer the possibility to manage efficiently both the living spaces and the human activities, and today they are increasingly present in our lives for what concerns the security, the energy management, the environmental control, etc. In educational spaces additional needs are involved, such as the management of teaching activities, libraries, exchange of information between teachers, students and parents. The modern design of school cannot avoid the use of intelligent networks (smart grids). Just consider, as an example, the energy management and the environmental control. The EPA (US Environmental Protection Agency) identifies as a top priority the control of air quality in school facilities because, as recent studies carried out in USA clearly show, the indoor pollution can reach levels even 100 times higher than the outside (outdoor) environment; the problem is particularly acute in schools, where a population of very young age spends a large number of hours per day. Recent studies show a significant increase in respiratory problems in schoolchildren (https://www.epa.gov/iaq-schools). The recent development of low-cost sensors able to monitor different air pollutants (solid, gaseous, but also organic), together with the development of platforms able to handle on line and in real time the information produced by the sensors, now allows the collection of statistically significant data regarding the concentration of air pollutants, in both space and time. These systems thus allow characterizing, besides the amount, also the causes of pollution, permitting to design interventions to mitigate the effects. The development of alternative energy sources is also dependent on smart grids; in structures such as public buildings (schools, but also hospitals, barracks, offices, museums etc.) the energy needs could be easily ensured by innovative small-scale solutions, such as micro-wind, solar, micro-hydric, etc., whose management is only possible through smart grids. The same could be extended to the medium scale of neighbourhoods and cities, where the proper use of ICT would allow, either directly or indirectly, significant economic savings for the community.

(II) Coming back to innovative teaching spaces, an interesting point is that the technologies described so far (environmental monitoring, network connections and data transfer, collection and analysis of information), are modular and easily accessible at low cost. This means that they could be conceived and developed by students within the same educational facilities where they are used. In other words, the design of these systems, including hardware as well as software, and development of ideas, could become a permanent laboratory where students learn how to focus on a problem (e.g. the presence of solid pollutants in the air), to seek a solution (search for sensors able to measure the pollution), design an intervention (i.e. conceive a measuring system and data saving/transfer), build a hardware system (for example a sensor platform), display the data (via construction of graphs on spreadsheets like EXCEL), understand the meaning of data, until the construction of thematic maps (GIS systems). Such teaching exercises are now easily achievable with low / very low cost components, can be understood easily from medium/high level school students, require concepts that are readily available in simple manuals or websites, and can be realized within the usual mathematics, physics or technical applications courses. From the pedagogical point of view, this type of education would develop in students the pleasure for the invention and the creation of objects, placing them within that community, or cultural international movement, known as the makers community. Furthermore it would develop in the student the direct perception of the link between concepts that would otherwise remain disconnected from each other, such as: a numerical datum (e.g. a concentration supplied by a sensor), its variation in time and its physical meaning (a rate of pollution), besides allowing students to understand directly their environment and finally respect it.

BIM (building information model)

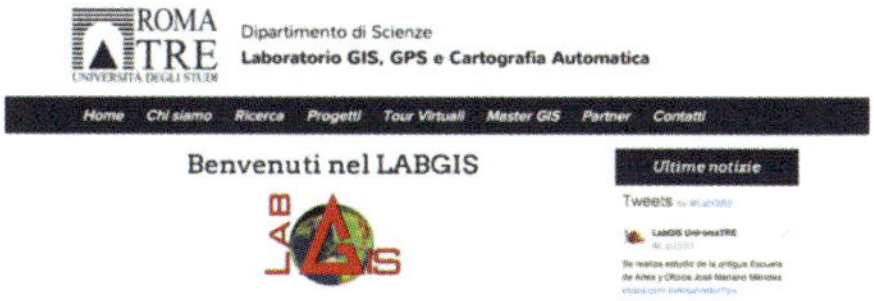

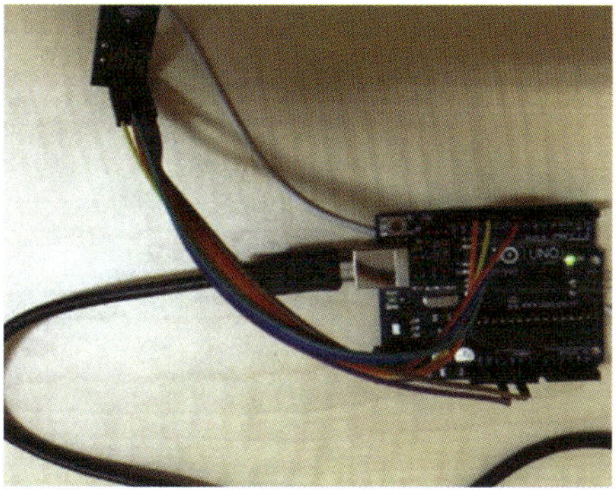

GIS Laboratory of Rome Tre University
(http://host.uniroma3.it/laboratori/)

2.3

The architecture technologically sustainable

Giovanni Mazzitelli*

> *"The sustainable development is not a fixed state of harmony, but rather a process of change in which the exploitation of resources, the direction of investments, the orientation of technological development, and institutional change are made consistent with future as well as present needs"*
>
> Brundtland Report, 1987

The Sustainability

Often we refer to the concept of sustainability in environmental terms, simplifying it with the meaning of *leave for future generations a planet in the same condition* in which we have found before a life spent on it. In fact this concept is broader and defines our relationship with science and technology. The equilibrium of tree macro systems is actually needed to be sustainable: *economical, ambiental and societal*. If these systems know how to communicate in a harmonious and respectful way we can create sustainable products and activities.

In all that science and technology are not actors, but tools. Tools in continuous evolution whose functionality is often limited by the will of one of the three actors, the economic system, whose *utilitarian faith* is based on short-sighted exploitation of the other two macro systems. This happen in the name of an impossible infinite growth, the limits of which are to be found in laws of physics, in the laws of thermodynamics, in the entropy (*A. R. Romeiro, H.N. Sá Earp, The Entropy Law and the impossibility of perpetual economic growth, arXiv:1309.2274v1 [physics.soc-ph]*).

Nothing is lost, nothing is created, everything is transformed... in nature there is no process that can really create something from nothing, there is always something or someone that is *exploited* or a place to which divert resources.

So on one hand the technology allows mankind to make more *efficient* use of natural processes, getting closer to a more sustainable economy, on the other science requires us strict rules to which humanity is subject, forcing to a *bio*-economy that assumes the paradigms of science as fundamental and inviolable (*N. G. Roegen – Bioeconomia. Verso un'altra economia ecologicamente e socialmente sostenibile, Bollati Boringhieri, 2003*.

The Technology

In this context what role absolves the technology?

* Giovanni Mazzitelli, INFN, National Institute for Nuclear Physics.

The human species has a great advantage than the other species that inhabit the earth: in addition to biological evolution, that through natural selection allows the development of genetic traits, ensuring the survival and competition with other species, the humanity can use in its favor also the technology "evolution".
Since the man invented the first tool, extending his physical skills and survival, has become much more efficient. The man, in fact, with an immediate gesture could pick up a stick from the ground by extending his arm to look in the bushes without being bitten by a snake, while the giraffe had to wait thousands of years to develop a longer neck.
A linear development has characterized our civilization for thousands of years. Great progress, such as the ability to cultivate, to use animals for heavy work, the wars, the birth of the city, have not fundamentally altered the planet.
But 200 years ago the first industrial revolution, he has accelerated our ability to be able to dominate the physical world (*E. Brynjolfsson, A. McAfee La nuova rivoluzione delle macchine, Feltrinelli, 2015)*, mainly because with the tools designed by us we were able to use unimaginable quantity of energy.
A typical nuclear power plant, which occupies the surface of a few hectares, is capable of producing the work of more than 1.3 million horses, while to power a city like Rome would serve about 2.3 million horses that works continuously.
Besides, we can store a wealth of information in a way unthinkable 20 years ago - a long novel is about 200 folders, one folder is approximately 1500 characters, equivalent to 300KByte of information, a disc from 1TByte, typically inside in our PC, contains therefore at least 33 million novels.
This is a huge opportunity that we have not only to learn how to take advantage of, but that have to handle by future generations. But, the technology is primarily a tool, an instrument at the service of man, a tool that young people need to understand, appreciate and dominate.
Its function it's to amplify our senses and to give us a chance to extend beyond those limits that the places require us, beyond the walls of a school, as well as beyond the standard methods of learning.

The architecture and learning spaces

What have to do sustainability and technology with the design of innovative learning spaces?
The architect, surely is the figure that is more then any others is in front of the challenge of having to imagine and, at best, anticipate one of the fundamental aspects of our future that are the spaces in which our life will take place. Also it is constantly forced to confront and mediate in the planning, making what is sustainability in its concept more complete, integrating environment, society and economy in a space that must be effective for today and for a future of tens of years.
A responsibility that takes even deeper meaning at the dawn of the fourth industrial revolution (*K. Schwab, The Fourth Industrial Revolution, World Economic Forum, 2016)*, where the machines will lead to a new social revolution, hardly imaginable today.
Hence, designing innovative learning spaces, means allow those who live in them to learn in a different way: an inclusive school which recognized *in a place for anyone a space for everybody*.
The new youth, probably, will not live in a world of skills and labors, as we know today, of defined roles, of tangible products and services and of a life spent to deepen their knowledge. In this plausible replacement with machines, now capable of learning and decide to use objects as man does, the challenge of young people will be to be able to *innovate and innovate them self* and to develop the *ability to learn*.

Hence, it changes the paradigm with which to build learning spaces, thinking not a place for the individual skills and specializations, but to a space for the growing in which people are able *to develop the identity, autonomy and skills, namely, learn to reflect on the experiences through exploration, observation and practice to confront each other* (*Programmazione educativa della scuola dell'infanzia, MIUR, 2007*).
This probably puts an end to the concept of static learning space, frontal, classic, requiring a change in approach by which until now have been designed and built classrooms, courtyards, libraries.
This opens the way to the design of dynamic spaces, where methods and contents to be learned are in constant change. Today, thanks to the results of the third industrial revolution, the contents are always - technologically - available, and our expertise is no longer in possession of the data, but in knowing how to transform this information into knowledge, in other words, in the awareness and understanding of the facts.

Technology is a great opportunity: it can be useful to create spaces that can allows young people to discuss on the technology giving them the knowledge needed to take the most advantage from the technology itself; It may be the tool for thinking sustainable places integrating environmental, social and commercial solutions in a timely manner; It may be a channel through which educate to the concept of sustainability, efficiency, unique milestone reachable due to limits of the laws of nature; It can be an instrument of expansion of our senses that remain the basis of any method of learning. It's up to us that the *technology*, the *speech on arts*, on *professions* as suggested by the etymology of the word, remains a tool at the service of humanity, rather than become a technocracy of the machines on our lives and in the spaces to be designed.

References

- N. G. Roegen (2003), *Bioeconomia. Verso un'altra economia ecologicamente e socialmente sostenibile*, Bollati Boringhieri, Torino.
- A. R. Romeiro, H.N. Sá Earp (2015), *The Entropy Law and the impossibility of perpetual economic growth*, Open Journal of Applied Sciences, 5, 641-650.
- K. Schwab (2016), *The Fourth Industrial Revolution, World Economic Forum*, ISBN: 1944835008.

2.4

School buildings requalification over adaption to standars

Antonello Monsù Scolaro*

The statistical surveys on existing school buildings have clearly demonstrated the poor condition and highlighted that several requalification interventions should be needed. In Italy, "they lack both a systemic approach and programmatic actions on public real estate, also missing the exchange of knowledge and skills between different fields, instead it is full of expensive and ineffective managements" (Losasso, 2015). The interactions among scientific expertises, binding for effective rehabilitation of school buildings, became complicates in this scenario.

In recent years, due to lack of maintenance, the conditions of school buildings are very heterogeneous. Generally, the 55% of the buildings was built before 1976 (the 4% before 1900; the 4% before 1920; the 8% before 1945; the 12% before 1960 and the 27% before 1975); the 32% was built after 1976, whereas data of 13% of buildings are missing. In order to verify the consistency, condition and functionality of school buildings, the law No 23, January 11 1996, 7th paragraph (even integrated by law No 431 August 8, 1996) established the civil registry of school buildings. The civil registry, in the first part, contains data about property, uses and age of the buildings, as well as the maintenances made over time; whereas compliance with safety standards and fire prevention are in the second part. The civil registry was updated progressively in 2010, 2012 and 2015: the last scenario is really discouraging.

In addition to the data in the table, the survey of the Ministry of education has verified both obstacles preventing acces to handicapped and ramps; paths; lifts and stairlift (71% no; 29% yes). Moreover, was made several maintenance works aimed to energy-saving improvements by double window panes; roof and envelop insulation; thermal plan zoning; solar panels (58% yes; 42% no). Previously, the Censis survey of May 31, 2014 (diary of transition) reported the outlooks of 2,600 heads school interviewed about actual conditions of school buildings. From the interviews it emerged that 36% of the buildings needed urgent maintenance works, whereas in 57% of cases it must finish already started works. However, despite the school buildings conditions weren't adequate to the current user's needs, only in 7% of cases the displacement to another location was required by school heads. Beyond the need for structural adaptations of school buildings, the data seem showing an ongoing "run-up useless to adapt to standards school buildings that will be out of standards from the next adaptation to standard " (Campagnoli, 2013). Therefore, it is necessary wondering if the measures taken have been effective in time or whether they have lost sight of other important urban, social and cultural issues.

* Antonello Monsù Scolaro, DADU, University of Sassari.

Indeed, in this scenario, in may 2015, it was written the first triennial national programming of interventions concerning two years period 2015-2017; it is part of a more general national plan of school construction. The expected interventions are over 6,000 totalling 3.7 billion euros. For the first time, at least in the planning documents, it seems that the adaptation to standards has been addressed in a more organic way than its actual complexity.
The funding should cover the structural safety renovations envisaging both anti-seismic retrofitting and accessibility improvements, as well as regular maintenances (painting and small restorations). It also considers the construction of new schools. Precisely, from the planning it emerges "#innovative schools. We'll design the school of the future". Perhaps, this slogan represents the real break with the past because it urges us to reflect on places of learning taking into account both quality and characteristics of the architectural space (in cognitive processes) by giving it more importance than to technical and regulatory requirements. In fact, while the introduction of ICT in the school has transformed the temporary dimension of the lesson in recent years, the physical space isn't changed; progressively, it emerges the need of adaptation to the new educational requirements.
In 2013, in Italy, the Ministry of Education formed a working group of architects and experts in the field of educational innovation in order to review the technical standards for school buildings; then, the topic of "innovative schools" turns out to be the result of that initiative. The new Guidelines of School Construction was approved the April 11, 2013. They contain many interesting points to provide an overview of the "school of the future", also thanks to the elements emerged from the questionnaires -according to a form "participated" in the definition of the objectives- addressed to heads school (active viewers, sometimes victims of ever-changing planet school). The Guidelines offer a virtuous and operating scenario that aims to be homogeneous (where possible) in a national level, focusing in particular on quality of the spaces of learning: classrooms and laboratories; individual and working group spaces; informal and relaxing spaces. Furthermore, paying attention to urban aspects -location, environmental quality and accessibility- the guidelines gives value to the quality of supplementary and ordinary spaces for the development of school activities (gymnasiums, laboratories, open spaces, etc). Particularly, for the first time, you may point out importance gave to the quality of indoor environments adopting of adequate technological systems; "health-friendly" materials; nonhazardous furniture. This shows an awareness on the quality of learning places, "that is not indifferent to learning processes" (Campagnoli, 2016). At the same time, pushing descriptive limits of guidelines, it is clear the intention to design the school as a "system of places" imagined by own users through an active participation in the design process, experiencing the sensations of its own body domiciled in space that they live (Pallasmaa, 2009).
In other words, if the school building was "perceived" and not just "used" by its inhabitants, it could once again become a social catalyst for its importance in the urban fabric. In conclusion, the standard adaptation might be an important opportunity to redevelop parts of city also lacking identity and social attractiveness (Faroldi, 2015), like an opportunity for a more articulated and complex regeneration process of the places of education, that should become an active and dynamic part of the social and urban fabric.

References

- Losasso, Ricerca tecnologica ed architetture dei servizi per la sanità e la formazione, in Technè 9/2015, pag.6-8
- Campagnoli G., 2016, Oltre le aule, Lulu.com Edizioni online
- Campagnoli G., 2013, Linee Guida per l'edilizia scolastica, un passo avanti?, http://www.educationduepuntozero.it/politiche-educative/linee-guida-l-edilizia-scolastica-passo-avanti-4074004890.shtml?refresh_ce-cp, access 12/06/2013
- Pallasmaa J., 2009, Gli occhi della pelle. L'architettura e i sensi, Jaca Book, Milano
- Faroldi E., Architetture per la salute e la formazione. Lineamenti e tendenze, in Technè 9/2015, pag.9-13

Experiences of research and project:

Learning as a therapeutic treatment

Josep Miàs*

The psychiatric centre of Arenys, Can Zariquiey, is a project for a young people mental health centre. This centre is quite unique in its genre because every treatment is based around the cultivation and cure of orchids. Indeed, just as the doctor takes care of the patient, learning how to care and cultivate and orchid functions in the same way and is a therapy itself. Caring for the orchids is a routine activity that requires great precision and helps the patients to generate a coherent, observant and disciplined attitude that improves their mental and physic state.

For this reason, the project assumes great relevance and centrality as the main focus is placed on the construction of a big orchid greenhouse that is placed adjacent to the main building.

The building is designed as a small intimate school where the young people can become specialists in the cultivation of the orchids. After the plants have grown the centre is then converted into an important flower market for the public. This gives importance to the work of the patients rewarding and acknowledging the result of their efforts.

Alongside with the greenhouse/workshop the centre, obviously, has other areas that include ateliers, conference rooms, common areas, bar and external space. All the spaces are designed as learning areas closely related to work and manual exercise.

This singular health centre works on a co-operative based concept and every decision and discussion becomes a collective and consensual act.

I believe that this founding decision reflects on the design and conception of the building and the decisions are taken consensually with the patients and users of the centre.

The new spaces are drawn considering the existing building. Some parts have been preserved, others have been altered and when needed new volumes where added. Each decision was made thinking in a smaller domestic scale designing the rooms as if they were living spaces of a house and choosing the furniture accordingly to further highlight this collective/family atmosphere.

It is a meticulous chirurgical work and the intervention respects and places a lot of attention to the existing context. The project valorises, preserves the existing qualities while adding new spaces and functions that respect the values and conditions of the existing context.

* Jose Miàs, MIAS Architects.

[1]. In 2015 Can Zariquiey Health Centre is finalist at the European Union Prize for Contemporary Architecture mies van der Rohe Award; in 2014 he was awarded at the Venice Biennale.

The greenhouse and its orchids become the new pulsing heart of the psychiatric centre and thus the decision to design this area in detail as a pivoting space for the centre and treatment of the patients. Chosen as a project for the 2016 Venice Biennale we decide to highlight the collective work and collaboration that brought the architectonic result, leaving the professional aspect aside. The presentation of the project, indeed, is a big collage of photos taken by the patients. There are photos of orchids, plants, pavements and furniture in which one can clearly see the hands of the patients that let us understand their effort and care of things. The images overlap, intersect and cut just as an architectural project layers necessities and technical details and joining these two sides one can arrive to a final shared result. What a pity that the organisers (architects) vertically placed the collage that in reality represented horizontal working table!

CAN Zariquiey, health Centre[1]

Author: Josep Miàs Gifre architect, Valentí Agustí, Psychiatrist
Office: MIAS Architects
Collaborators: Adriana Porta, Silvia Brandi, Fausto Raposo, Daniel Montes, Nuno Marques, Sebasti de Iruarrizaga, Alfonso Ab, Christian Giovanetti, Giovanni Galdieri, Gabriele Mura, Laura Pomesano, Marc Subirana, Xavier Bas Structural
Engineering: Anabel L Civil, soil and survey: Josep Masachs (PROISOTEC)
Program: Healt Center
Location: Arenys de Munt, Spain
End of works: 2013

Project Photomerge

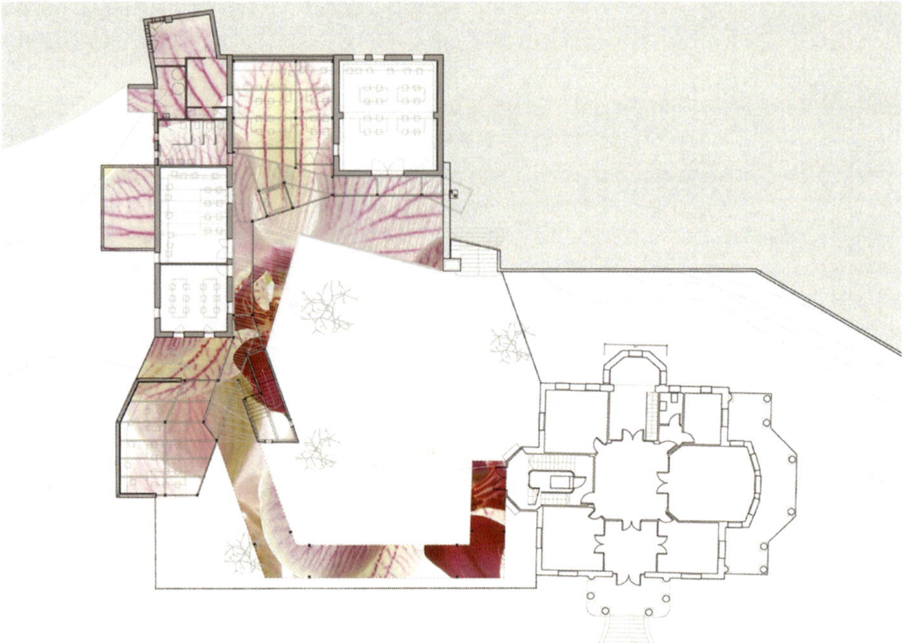

General Plan

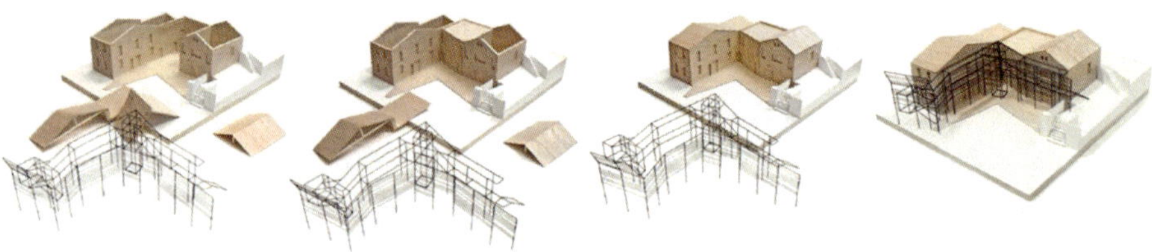

The corridor space, the ateliers and the patio

Dedicated Polyvalence

Patrick Fransen*

Schools are an important part of our knowledge in my practice, loyal to a strong pedagogical concept. NOAHH believes in the social aspects of the architecture and urbanism and how this can change the relations of the users and their way of communicating to each other. On the other hand we focus to the needs of the child and provide them a save and well organized environment. Only decennia ago and even still (!) most schools are based on the pedagogy of frontal education with closed classrooms. In current times where information is available everywhere a more social interactive environment becomes more important. Schools with more focus on individual- and group-education. This attitude is shown by some of my projects.

Example 1: Extended school Savio-Hilmare, Hillegom (NL)

A traditional school is based on a corridor typology with large classrooms at the side. In this school we enlarged the corridor and programmed it with workplaces for more informal activities, an open stage, meeting rooms, voids and also more sheltered places where students can withdraw. The classrooms are open and transparent and can be connected with the more informal areas by opening their large sliding doors.

The new school building for the schools Savio and Hilmare will consist of an ensemble of old and new. The characteristic gymnasium - part of the original building - will continue to exist and be transformed into a multifunctional hall. The new building is conspicuous by its overhanging roof, which unites the different users. Open-plan floors combined with the large transparent facade create a bright and airy school. The materialization is very neutral in its base and is decorated with wood finishes in wardrobes, sliding walls, stairways and closets.

This school with its neutral base and materialization is also meant for children with autistic spectrum disorder. Good acoustics, a reluctant and quiet environment and furniture are important conditions to facilitate this small school where children should feel easily comfortable. We discovered that the more informal area is very important, not only because it can be used for individual education but also to accommodate space for children to feel at home and to withdraw.

* Patrick Fransen, NOAHH – Network Oriented Architecture.

Extension of the Savio-Hilmare School, hillegom (NL)

Author: NOAHH - Network Oriented Architecture
Team: Patrick Fransen, Mick Madder, Ilse Bakker, Bilal Karaburun, Kristina Kosic, Paul van Dijk (Studio Deep Blue) Anouk Aerts
Surface: 1970 sq m
Client: Aloysius Stichting
Place: Hillegom
End of works: 2016

NOAHH: extension of the hillegom NL school, external 2015-2016

NOAHH: extension of hillegom NL school, interior 2015-2016

Example 2: 't Slingertouw Waterrijk Eindhoven (NL)

This school shows how it became a successful neighbourhood centre and innovative learning space. The main idea of the school is conceived on the organizational structure of a market hall. In a market hall you can find particular places for buying bread, fish and vegetables and particular places for drinking coffee. In this school students no longer have their own classroom but they are nomads in their own school-compartment. They share dedicated spaces for learning mathematics, language, crafts, drama. Students learn to work together and individually. Just as at the market hall, there is also a neighbourhood-café for students and parents.

Waterrijk primary school is not a self-contained building, it is linked to and incorporated into a residential block in a more comprehensive housing plan also designed by us. The upper floors accommodate a part of the extended school programme. If pupil numbers decline, this section can be converted for residential use after all. A neighbourhood centre with a café-like meeting-place is also part of the school programme, so that school and neighbourhood activities can share the same facilities.

The major part of the school is sunken below ground level and provided with a constant stream of daylight through inner courtyards (each of them positioned between two classrooms) and through elongated skylights above the more informal learning areas. Strips of classrooms are interrupted by these more informal areas: the learning streets. The learning streets are designed entirely to serve educational needs. They contain sunken sitting areas, conversation pits, 'help desks', small theatre-like enclosed areas, small kitchens and other facilities for all to use. The concertina doors of the classrooms can be opened up to create a single continuous learning landscape. The playground on the roof is also accessible to children from the surrounding neighbourhood. This school shows how a school can become not only an innovative learning space but also fully integrated in the neighbourhood. This is based on the idea of the school itself with its motto: "playing, integrating and learning."

Waterrijk Neigbourhood Centre t Slingertouw,Eindhoven (NL)

Author: Architecturstudio HH (Herman Hertzberger) with Patrick Fransen NOAHH - Network Oriented Architecture
Team: Jos Hafweeg, Reijer Pielkenrood, Francine van Loon, Zina Al Ednay, Carloes Pollemans
Surface: 6090 sq m
Client: SKPO Eindhovem
Location: Eindhovem
End of works: 2011

NOAHH / AHH: Waterrijk Neighborhood Center t 'Slingertouw, Eindhoven NL, interior

NOAHH / AHH: Waterrijk Neighborhood Center t 'Slingertouw, Eindhoven NL, interior

NOAHH / AHH: Waterrijk Neighborhood Center t 'Slingertouw, Eindhoven NL, interior

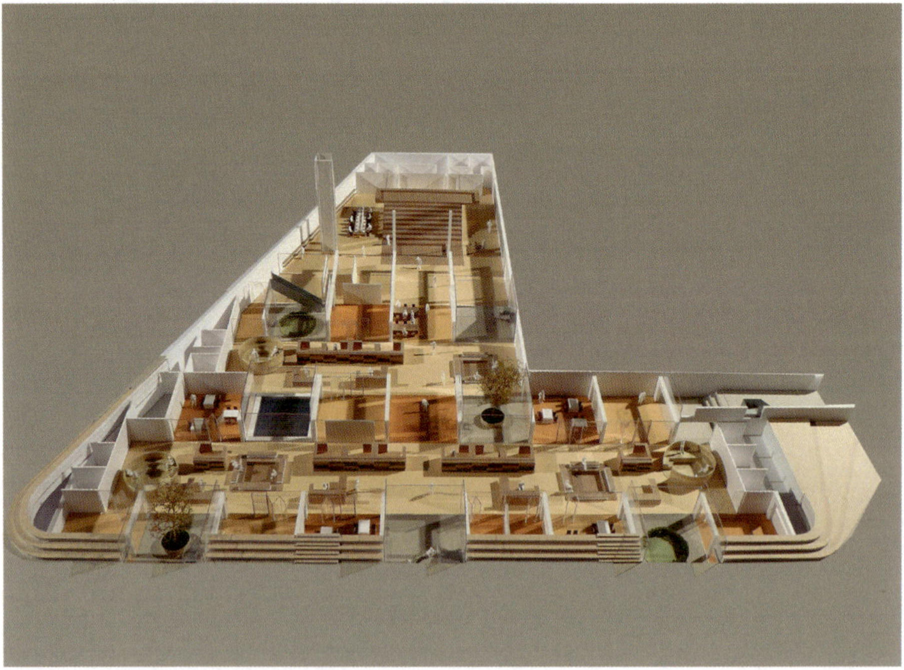

NOAHH / AHH: Waterrijk Neighborhood Center t 'Slingertouw, Eindhoven NL, model of the open plan

NOAHH / AHH: Waterrijk Neighborhood Center t 'Slingertouw, Eindhoven NL, interior

Belgium

Like in many countries the schools in Belgium are often very outdated and the pedagogical attitude is often still based on frontal education principle. By winning some competitions we could test our attitude here to debate with the schools and to start a transformation process from school to innovative and inspiring work- and learning space: with success. A liberal open floor plan is not always needed. But to strive for interaction by grouping functions and classes around communal and common functions already leads to a more integrated and thus innovative learning space: *"It is not the classroom that matters but the dedicated and programmed polyvalent space."*

Example 3: Science building, Sint Jozef Institute – Geel (B)

The building is conceived as a neutral structure with the greatest possible flexibility to plan layout. The building is strategically integrated into the fabric of the other school buildings and also aims to bring an improvement in the use of existing buildings. The flexible floor plan of 40x44 meters makes the building can respond to the evolving pedagogical understanding of the school. The large dimension brings in the need for patios and voids. Very unusual in Belgium but that was exactly what was needed to serve interaction between two levels.

Special attention is given to improve the quality of public space. The first floor of the new building is to be understood as a large roof (a giant umbrella) over the outdoor area. This constitutes into the most important gesture of the design: the school as a place to find shelter.
This strategy has successfully be repeated in some other schools for the same client 'KoGeKa'.

Science building, St Jozef Institute - Geel (B)

Author: Patrick Fransen NOAHH – Network Oriented Architecture
Collaborators: Studio Plus Architecten bvba
Team: Patrick Fransen, Ilse Bakker, Freke Schalken, Tim Loesters, Paul van Dijk, Henk de Weijer, Onne Walsmit, Loes Thijssen, Reijer Pielkenrood
Surface: 2800 mq
Client: DBFM Scholen Van Morgen - AG Real Estate COPiD - Katholiek Onderwijs Geel_Kasterlee (KoGeKa)
Location: Geel
End of works: 2017

NOAHH Science building, St Jozef Institute - Geel (B) project area

NOAHH Science building, St Jozef Institute - Geel (B) project area

NOAHH Science building, St Jozef Institute - Geel (B) exterior

NOAHH Science building, St Jozef Institute - Geel (B) interior

3
Devices for learning: reading and interpretation

A substantial difference exists between the quantity of urban allocations and the tangible capacity they have to accommodate the project requirements useful for every single inhabitant. To intervene in designing urban space means to responsibly reconsider the role space actually covers in the dynamics of the processes of use and appropriation of the city, meant as personal courses of action and knowledge.
Nowadays, these needs are often fitted into spatial elements easy to recognise, sometimes completely clashing with the context, that denounce their presence as urban objects, metaphors of ideal spaces closer to dreamlike worlds than real ones.
They are the playgrounds where it is the architecture itself that learns from play, to the point of becoming inconstant and interactive, in which 'doing' turns into a landscape.

#inhabitant #appropriation of the playground-city #flexible and modular space

Focus Group dedicated to learning devices: readings and interpretations.

85

3.1

Learning devices: in the *wake of a mutation*

Paolo Calidoni*

School devices and related architectures that we inherited from the 20th century give testimony of a transformation that on the Millennium ridge has increasingly assumed the character of a radical change that challenges research, design and practices.
Schools in modernity, as Foucault (1975) masterfully highlighted, are configured as institutions of national states, functional to standardization and social selection, based on the use of books as the main teaching tool.
The school architectures, therefore, have interpreted, shaped and made possible such functions with impressive buildings and homogeneous/standard classrooms, panoptically controllable, where the board and the chair -with schoolbooks and logbook- were the focus, the focal point for the delivery of ' teaching in homogeneous groups (classes) of pupils in uniform, still at their desks, evaluated and selected on compliance with the model of a subject / citizen / worker manual or intellectual form.
Most of the historic buildings of past centuries, still working, has this structure that today still constitutes the archetype of the school in the common idea.
The idea of school in the 20th century has shifted from teaching to learning, from the instructive function and selective focus to the social and educational care, from control to exploit the the pupils/ students potential.
Schools and teaching practices have become more complex and have taken on different settings (Ferrari, 2015), less standardized than according to a unique educational and transmission model, but still with a strong national character and centrality of the book (and more and more of the library).
Beginning with the Montessori 'Casa dei Bambini', tailored for childhood growth (Fig.1), the interaction between indoor and outdoor spaces -particularly for rural-school classes in the italian rationalist architectures- (Fig.2); up to the generalized diffusion of mobile desks (Fig.3) for individual and group activities, to 'open-spaces', laboratories and workshops, especially in the early grades of education,.
The dialectic tension between teaching and learning and the research and carrying on of new settlements has characterized and characterizes the daily life of teaching practices and has reverberated and materialized in setting and use of the available spaces.
We find, then, old halls of historical buildings where the chair is in the corner and the '60 desks are aggregated so as to favor the activities for groups of pupils (Fig.4); but also formerly open-spaces divided in classrooms or laboratories with masonry walls and where the desks and pupils are always aligned in rows and columns, facing the chair and LIM (Calidoni, 2016).

* Paolo Calidoni, former teacher, school manager and chair of the Committee for School System Evaluation of Trento Province, is full professor of didactis and educational research at the Parma University.

We are witnessing attempts to 'domestication' (Appadurai, 1996) of the new technologies to the 'forme scolaire' (Vincent, 1994) consolidated over time, as well as alternative and creative use of the available spaces.
These readjustments, gradually more and more widespread and massive, highlight the search for equilibrium points in a dynamic of ongoing transformation that sounds so radical to appear more as mutation rather than a mere evolution, a jump, a paradigm shift generated by the accumulation of factors that overtake the modernity and next to the point of catastrophe, in the technical sense of the word and not a pessimist one (Thom, 1980; Tonietti, 2002).
In fact, the features defining the archetypal 'forme scolaire' of modernity are 'fallen' or replaced by others with an implicit, tacit and not always conscious change in the type of device used.
First, from standardizing and selecting institution, the school has become, at least in the educational and official rhetoric, a right for all which implies -especially in Italy- that the group of students are highly heterogeneous (although still are called 'classes ') and they share the same classroom; but a collective teaching for everyone is disfunctional because, as already recalled Don Milani with the boys of Barbiana School (1967), "there is nothing that is more unjust than deal equally between unequal ".
The 'inclusive' function of the school (Booth-Ainscow, 2008) is considered essential for its quality, therefore only one classroom for each grade becomes insufficient, inadequate and other spaces for socialization and for the diversification of the educational activity become necessary, and not merely incidental.
Complementary to the inclusive function of the school is the *inte (g) ration* in the social, urban and environmental, or rather more interaction between the school and other learning opportunities and spaces, to overcome stiffness and institutional separations and increase flexibility, osmosis and hybridisation such as those that occur, for example, when the school is also configured as civic-center, or when more than one language is spoken and have equal status as the national or mother one (UNESCO, 2012).
Last but not least, even more macroscopic, is the mutation of the transmission and processing tools of information and knowledge that sees the book alongside digital technologies in the network, inside and outside the school, with a osmosis that thins or breaks down boundaries and specificity (Pireddu, 2014) and suggest incipient radical unpredictable changes.
In short, the archetypal school paradigm, established and delivered until the 20th century, in the last decades has been upset by the change of its supporting pillars, but it is still available and often dominant while a new one is still researching his semantics, syntax and grammar.
Today, the design of learning spaces is challenged by the paradigm shift in which we are involved. Also the new 'digital native generations' (Prensky, 2001) and mainly who is committed to design and test new solutions, must be able to ride the wave of the ongoing transition: 'antifragile ' (Taleb, 2013; Blecic-Cecchini, 2016).
In this perspective the systemic planning illusions need to be overcome to allow evolutionary, transformative, open and flexible processes, to take space capable of empowering of potential through strategic guidance and 'visionary' solutions able to generate possibilities for further action.
The 'mechanical' teaching-learning device of modernity is leaving more and more space to an organismic, ecological and networked vision, centered on an integrated model of the human mind, including biology, culture and technology (Piredda - Di Francesco, 2012).

Spaces for learning and educational and didactic action are then getting reconfigurated in a new perspective (Dumont-Instance-Benavides, 2010; CELE, 201; Biondi-Borri-Tosi, 2016; Fondazione per l'architettura, 2016) (Fig. 5).

The methods, timing and quality of the process will also depend on the orientation and support that research (interdisciplinary), political and professionals will be able to offer.

References

- Appadurai A. (1996), *Modernity at Large*, University of Minneapolis Press, Minneapolis.
- Biondi G., Borri S., Tosi L. (a cura di) (2016), *Dall'aula all'ambiente di apprendimento*, Altralinea, Firenze.
- Blecic I., Cecchini A. (2016), *Verso una pianificazione antifragile. Come pensare al futuro senza prevederlo*, Angeli, Milano.
- Booth T., Ainscow M. (2008), *L'index per l'inclusione. Promuovere l'apprendimento e la partecipazione nella scuola*, Erickson, Trento.
- Calidoni P. (2016), *Immagini dalle aule. La perdurante predominanza del modello didattico 'uno-tanti'*, in Scuola Democratica 1/2016, Il Mulino, Bologna.
- Centre for Effective Learning Environments (2011), *Designing for Education: Compendium of Exemplary Educational facilities*, OCSE, Paris.
- Dumont H., Instance D., Benavides F. (eds) (2010), *The Nature of Learning. Using research to inspire practice*, OECD, Paris
- Ferrari M. (a cura di) (2015), *Di ogni ordine e grado. L'architettura della scuola*, Rubbettino, Soveria Mannelli
- Fondazione per l'Architettura, Torino (2016), *Lo spazio nelle scuola innovativa – La scuola secondonoi*,http://www.fondazioneperlarchitettura.it/wpcontent/uploads/2017/03/F_ArchScuole_LT.pdf (26/03/2017)
- Foucault M. (1976), *Surveiller et punir. Naissance de la prison*, Gallimard, Paris.
- Piredda G, Di Francesco M. (2012), *La mente estesa. Cervelli, tecnologie e l'enigma della natura umana*, Mondadori, Milano.
- Pireddu M. (2014), *Social Learning. Le forme comunicative dell'apprendimento*, Guerini Scientifica, Milano.
- Prensky M. (2001), *Digital Natives, Digital Immigrants*, in On the Horizon 5/2001, MCB University Press, Bradford.
- Scuola di Barbiana (1967), *Lettera a una professoressa*, Libreria Editrice Fiorentina, Firenze.
- Taleb N.N. (2013), *Antifragile. Prosperare nel disordine*, Il Saggiatore, Milano.
- Thom R. (1980), *Parabole e catastrofi*, Il Saggiatore, Milano.
- Tonietti T. (2002), *Catastrofi. Il preludio della complessità*, Dedalo, Bari.
- UNESCO (2012), *Guidelines on Intercultural Education*, UNESCO, Paris.
- Vincent G. (1994), L'éducation *prisonnière de la forme scolaire? Scolarisation et socialisation dans le societies industrielles*, Presses Universitaire de Lyon, Lyon.

1| Spaces for education according to the Montessori method (Ferrari, 2015, p. 44)
2| Fifth Milan Triennial, Examples of rural classis (Archivi della Triennale di Milano; Ferrari, 2015, p. 54)

3| Achille e Piergiacomo Castiglioni, Luigi Caccia Dominioni, *Chair & desk T12*, 1960 (Fondazione Achille Castiglioni, Milano; Ferrari, 2015, p. 76)
4| Domestication of classroom setup (Calidoni, 2016, p. 33)

5| *"Manifesto for the third millennium educational spaces"* (INDIRE / National Institute for Documentation, Innovation and Educational Research, http://www.indire.it/wp-content/uploads/2016/03/ARC-1603-Manifesto-Italiano_LOW.pdf - 26/03/2017)

3.2

The school as a public space [and the city as a classroom]

Francesca Arras, Elisa Ghisu,
Paola Idini, Valentina Talu*

We contemplated at length what it meant to design truly *innovative* learning environments, focusing on what contribution the spatial project could offer to build truly "healthy" cities. Beyond the current definitions and guidelines used to define the word healthy, we would like to assign a more functional meaning. For us, a "healthy" city is a city that is both "usable and used" by all its inhabitants in a full, inclusive, innovative and secure manner.

A malleable and transformable urban context is not simply a healthy environment, but also as a tool that enables its inhabitants to be promoters of their own health. We are firmly convinced that it is possible and useful to go beyond the major urban regeneration projects and new constructions to instead focus on smaller interventions, editable and implementable at a neighborhood scale[1], while choosing to work with pre-existing errors and imperfections. The goal is not to create technologically advanced cities or schools, but to use the *appropriate* technological solutions (Oosterlaken, 2012, 2015), not always related to innovation (in the strict sense) - but to what innovation and technology provides, making it possible to "heal" existing spaces and extending the rights to the city to everyone, creating opportunities that promote "urban capabilities" (Talu, 2013, 2014) for all categories of inhabitants who find the city hostile and inaccessible[2].

We must investigate which technologies are indeed *appropriate*, or rather able to improve the usage and alteration of the inhabitants; especially considering the general tendency to operate on a very tight budget and often on existing schools characterized by various disadvantages and marginalization, from both a functional and social point of view, and accompanied by urban, environmental and architectural deterioration.

* Tamalacà Team.

[1]. For some thoughts on this subject, please refer to A. Cecchini, Talu V. (2012), "Contro la sparizione della città", in *Il calendario del popolo*, 757.

[2]. Those who are systematically excluded from the city project in virtue of their bodies divergence - in terms of shapes, abilities and lifestyles - from the body of a standard inhabitant, on which the design, construction, organization and government of the city is based, in spite of its obvious inability to represent the needs and aspirations of a multifaceted community. (Talu, 2014)

[3]. The reflection on the city as a shared and contended space is described in the writings of G. Paba (1998, 2002, 2010, 2012).

The assumption is that the city, as a whole, is basically a shared space[3], and in this sense, so are its Schools. By virtue of this awareness we believe it is both possible and desirable to construct (or better: to make possible) shared/sharable spaces within the various sections and between the different functions, but not in a way as to pose an imposition.
Furthermore, the possibility of sharing some of the living functions should not be disregarded, redirecting the perception and the project towards dynamics and requirements that trigger new ways of engagement, activism, care and sharing, opening the learning spaces to the city and vice versa, with the idea that "a space becomes public when it tests the coexistence, when addressing the problems and conflicts that arise from sharing the space" (Crosta, 2000).
In this way, not only the school can become a public space but also the city can become a classroom (McLuhan, 1977), because "the city is in itself an environmental education, and can be used to provide one, whether we are thinking of learning *through* the city, learning *about* the city, learning to *use* the city, to *control* the city or to *change* the city "(Ward, 1977).
This macro-topic has extraordinary depth, both physically and functionally, but also its meaning and social/cultural sense, favor the onset of a coherent and systematized ensemble of design and planning resources, using the same innovating tools for the project, both at an urban and architectural scale.
We believe that the school, finally rethought and rebuilt as a true public space, can be an *enabling* factor and device for many reasons:

- in prospect , evenly distributed on a regional level with a capillary diffusion, the creation of public spaces throughout the entire city gives a sense of quality and proximity to all of its inhabitants;
- the Community-School's ability to "build a community" and "build a network";
- the future possibility to make the students, families, and the entire staff at the school, not just exclusive recipients of the actions, but the agents of change themselves, in spatial terms and more, to become the co-authors of a truly inclusive and empowering community, rebuilding social forms out of the modern paradigm, rules, values, conflicts and reasons (because if it is true that on the one hand we have the inclusion / exclusion devices, apparatuses that urbanism has set in motion for a selective access, on the other hand there are also relational devices that seek to articulate the space and propose a system of shared spaces)[4].

As Tamalacà, we experienced this in projects to promote home-school pathways, thought of as an opportunity and instrument to reclaim the basic right of children to move autonomously, with some projects re-claiming small marginal spaces, developed following the approach of 'Tactical Urbanism' (Lydon, Garcia, 2015), with the idea being that it is necessary to experiment with new approaches and refine *innovative* and more effective tools for a more conscious and responsible involvement and activation of the inhabitants - especially those most disadvantaged[5], while being vigilant of both the participants as well as observers in the process.

---

4. The reference is related to the theme of *totalizing institutions* that have produced the discipline and control of bodies, standardization and functionalization as a dominant criteria, inhibiting the constructive and constitutive forces, producing subtraction of possibilities and potential rather than the addition and care, well evident in the reflection of Foucault (1975, among others) and Paul Rabinow (in Sandercock, 2004), which analyze the role of professionals controlling and regulating bodies in space.
5. Especially the most excluded groups, because they are, for us, more than the others 'design inhabitants'. For more details see in addition to Wolff, De-Shalit (2007), Talu (2013, 2014), Arras et al. (2013).

The following describes, via images, some interventions made by Tamalacà that follow this direction, in a tactical perspective, proceeding in a stepwise manner: from the first spatial actions through to the activation of transformational devices, chosen by working with the concept of a threshold for building stimulating opportunities, towards the construction of shared/sharable spaces, breaking the fences and boundaries, acting on the ambiguous transitions, invading the *in-between* and neighborhood spaces, creating a network.[6]

References

- Agamben G. (2005), *Profanazioni*, Nottetempo, Roma.
- Amin A., Thrift N., (2005), *Città. Ripensare la dimensione urbana*, Il Mulino, Bologna.
- Arras F., Ghisu E., Idini P., Talu V. (2013b), '*TaMaLaCà - Tutta Mia La Città. 'Suffragette' dei diritti urbani negati'*, Atti della XVI Conferenza Nazionale SIU *Urbanistica per una diversa crescita*, Napoli.
- Cecchini A., Talu V. (2012), 'Contro la sparizione della città', in *Il Calendario del Popolo*, 757.
- Crosta P. L. (2000), "Società e territorio, al plurale. Lo spazio pubblico – quale bene pubblico – come esito eventuale dell'interazione sociale", *Foedus*, 1.
- Foucault M. (1975), *Surveiller et punir: Naissance de la prison,* Gallimard, Paris, trad. It. (1976) *Sorvegliare e punire: la nascita della prigione,* Einaudi, Torino.
- Hertzberger H. (2008), *Space and Learning*, 010 Publischers, Rotterdam.
- Lefaivre L., De Roode I., (2002), *Aldo van Eyck. Playgrounds*, NAi Publishers, Rotterdam.
- Lydon M., Garcia M., (2015), *Tactical Urbanism:Short-term Action for Long-term Change*, Island Press, Washington DC.
- McLuhan M., Hutchon K., McLuhan E., (1977), *City As Classroom: Understanding Language and Media*, Book Society of Canada, Agincourt, Ontario, trad it. Lorenzini A. (a cura di), (1984), *La città come aula. Per capire il linguaggio e i media*, Armando Editore, Roma.
- Paba G., (1998), *Luoghi comuni. La città come laboratorio di progetti collettivi*, Franco Angeli, Milano.
- Paba G. (2010), *Corpi urbani. Differenze, interazioni, politiche*, Franco Angeli, Milano.
- Oosterlaken I. (2015), *Technology and human development*, Routledge, Abingdon, UK.
- Oosterlaken, I., Grimshaw, D.J., & Janssen P., (2012), "Marrying the Capability Approach, Appropriate Technology and STS: The Case of Podcasting Devices in Zimbabwe", in Oosterlaken I., Van den Hoven J. (Eds.), *The Capability Approach, Technology and Design*, Springer, Berlin.
- Sandercock L. (1998), *Towards cosmopolis: planning for multicultural cities*, Wiley, New York, trad. It (2004), *Verso Cosmopolis*: Città multiculturali e pianificazione urbana, Dedalo, Bari.
- Talu V., (2013), 'Qualità della vita urbana e approccio delle capacità', in *Archivio di Studi Urbani e Regionali*, FrancoAngeli, Milano.
- Talu V. (2014), *Qualità della vita urbana e approccio delle capacità. Perché e come promuovere le capacità urbane degli abitanti più svantaggati*, Franco Angeli, Milano.
- Wolff J., De-Shalit A., (2007), *Disadvantage*, Oxford University Press, Oxford.
- Ward C. (1977), *The child in the city*, Architectural Press, Princeton, trad it. (2000) *Il bambino e la città. Crescere in ambiente urbano*, L'ancora del Mediterraneo, Napoli.

[6]. It seems useful to think about the theme of 'threshold', and in general on the evolution of the concept of intermediary space - "in between" - reviewed by TEAM 10 onwards [especially Van Eyck, Hertzberger] because some open space pertaining to certain School specific contexts may be a tactical element, connecting and combining the urban fabric, restoring value (connection between public land and private land) and may also contribute towards a social nucleus and collective activities. This possibility linked to the creation of hybrid forms of *shared spaces, community-dwelling spaces/hub*, can be a strategic asset because it fosters mechanisms for social control and simplifies the management and maintenance procedures, transforming them into a collective responsibility, enticing the movement and passage of people, exchange and socialization, recreating spontaneously, from the behavior of the beneficiaries, a sense of community and participation.

1 | First spatial actions: activation devices (made with low-cost and playful tools). Sequence of images of the triggered actions for *MicroBodiesInMicroUrbanSpaces* project [MBiMUS]: "Every space is playable!" [Source: photographic material of the authors]

2 | It is precisely the playful aspect of the urban choreography by *MBiMUS* that helped implicitly reverse the perceptions of a city that appears to be controlled and blocked, activating a reflective process on land use (for playing and rest) due to the parked cars occupying the surrounding area of the school of San Donato in Sassari, desacralizing (Agamben, 2005) and emancipating from the ordinary everyday logic, linking the various and numerous multidimensional constructions that interlace the physical spaces and subjects (Crosta, 2000). [Source: photographic material of the authors]

3 | The evolution of the reflective process, post MBiMUS intervention, regarding a temporary installation in the parking-square in front of the school of San Donato in Sassari thanks to the release of two car stalls (made with arch. Fabrizio Pusceddu, Sabina Selli, Simone Censi). [Source: photographic material of the authors]

4

4 | Actions and projects aimed towards building shared/sharable spaces: the *PortaColori* project of Monte Rosello district, transformed an abandoned field into a colorful and safe open space, used by the school and surrounding neighborhood. [Source: photographic material of the authors]

5 | The images tell the story of 'Il Giardino che non c'è[ra]', and the relationship between the via Gorizia primary school of Sassari and the seniors residence 'Casa Serena' in the management of small regained space, and the perspective networking of interventions in the school courtyards of the city of Sassari. [Source: photographic material of the authors]

6 | The images tell the story of 'Il Giardino che non c'è[ra]', and the relationship between the via Gorizia primary school of Sassari and the seniors residence 'Casa Serena' in the management of small regained space, and the perspective networking of interventions in the school courtyards of the city of Sassari. [Source: photographic material of the authors]

3.3

University and school in the children's university experiment

Silvia Serreli*

University and school as agents of change

In urban contexts that are continuously subjected to change and new demand for cultural transformation, the process to educate inhabitants has a crucial role and needs to open up to a variety of experiments, including unusual ones. How can the university and school contribute to giving a different meaning to the elements that go to make up *living*, such as urban sociability, sense of belonging and quality of physical space?

Strategic city themes, like the quality of life and urban safety, climate change and emergencies, individualism and solidarity, require us to overcome the supremacy of fragmented knowledge "to make room for a way of knowing able to grasp objects in their contexts, in their entirety (...), we need to teach methods that will allow the mutual relations and reciprocal influences to be grasped between the parts and the whole in a complex world" (Morin, 2001).

Sassari University's cultural project "the Children's University" (UBU)[1] is a learning environment that reveals these themes and creates the conditions for acquiring transversal skills, nurturing critical thought and the ability to work in inter-generational teams and to involve others in tasks that may create a consensual dimension. The central role of children in the knowledge process, as different authors emphasise, is clearly highlighted by *student-centred learning* approaches (Kember & Kwan, 2000; Kember, 2008; Fabbri & Melacarne, 2015; Lisimberti & Montalbetti; 2016, Dettori, 2017)[2]. The activities UBU proposes do not aim to extend mastery of the knowledge of subjects proposed by the school but to make all the participants' (teachers, schoolchildren and students at university and at the Third Age University, parents and citizens) *learning to learn* capacities emerge (Morin, 2014), as Sisti & Torrisi emphasise (2016). This is one of the basic skills referred to in the *European Key Competences Framework* (European Commission 2006) but not always contained in current educational syllabuses.

* Silvia Serreli, DADU, University of Sassari.

1. An inter-departmental research centre that the Department of Architecture, Design & Urban Planning shares with Sassari University's Department of History, the Science of Man & Education. A project conceived by A. Cecchini and coordinated by S. Serreli and F. Dettori with the collaboration of lecturers and co-lecturers at the DADU_UNISS. It was set up and approved by two department councils and the academic senate in 2015. Sassari University's cultural project "Children's University" (UBU_"Università dei Bambini dell'Università di Sassari") aims to create permanent research, educational and social activities on themes linked with the right to the city and to life/learning environments. UBU activities address children first of all, without forgetting activities for teachers, parents, decision-makers and all citizens.
http://universitabambini.blogspot.it/p/la-scuola.html

2. Lisimberti & Montalbetti emphasise that teacher-oriented trends "do not encourage the individual and the community on the whole to invest in didactic innovation". Kember & Kwan 2000, Kember 2008, Fabbri & Melacarne 2015; quotations from these authors.

The perspective that steers UBU research is based on the possibilities of opening up "educational pathways" in both institutions where creative and experimental practices attempt a critical discussion on themes about which we need to be responsibly educated: freedom and democracy, ethics and the sense of otherness, progress and climate change, aesthetics and the role of art, evolution and the sense of history. UBU acts as an interpretative instrument of the city and its inhabitants, beginning with the youngest, who, thanks to their innate receptiveness towards innovative aspects, occupy the role of agents of change that operate not only via formal models but also by way of invention, reflection and learning during the course of the action. The "Children's University" explores transversal experiences between the two institutions that are able to make some of the urban life values converge which are arising from the many inter-generational views, the stories of individual development and different learning strategies. We can therefore consider UBU as the enthusiasm of the institutions (school and university) to create a common pathway by deftly responding to the educational demand and needs left latent or neglected by the traditional training on offer, often founded solely on the sort of teaching that is imparted. Edgar Morin (2015) associates these approaches, which currently prove to be rooted in the educational process, with the outcomes of the "disjunctive paradigm": in disjunction children are often forced to learn history, geography, chemistry, physics, mathematics, etc. in isolated categories. The necessity to avoid some strategies to separate kinds of knowledge that permeate the school and university has favoured a different "ecology of action" in UBU's activities, to quote Morin again, continuously stimulating a dialogue between the different subjects whose actions tend "to evade the intentions and will of their principal actor to enter into a game of interaction and retroaction with the (social or natural) environment that may modify its course, sometimes to the point of changing it" (Morin, 2015 p. 31).
The capacity to face the complexity and multidimensional nature of *living* in which unpredictability and uncertainty are constitutive elements forms the background to UBU programmes. This aim encourages a change of direction that involves each "learner", child-inhabitant or adult-inhabitant, in constructing a liveable city: "to favour the transition from motivation of a utilitarian type to more intrinsic motivation, which has the student engaged in a rewarding and increasingly autonomous educational route, of which he can take charge and thus be responsible for his own learning results" (Sisti & Torrisi, 2016).
In this sense the Children's University contributes to innovating those governing processes of the city that are most strategic for orienting the future and innovation, investigating the concepts of *lifelong learning* and *augmented learning* (Torrisi, 2013; Sisti & Torrisi, 2016) and therefore the development of awareness in citizens via different levels of education and of motivation to take care of their territory and, more generally, the spaces in which they live.

Educational experiments and action contexts

The contexts where the Children's University is active are not recognised as the outcome of negotiated processes of collective construction seeking consensus on the principles of education and citizenship, but are rather the result of educational, artistic and ludic activities in which many subjects have shared the unexpected effects of social interaction. The results of the different initiatives are illustrated in the reflections in the book "Città e formazione. Esperienze tra Urbanistica e Didattica" by Silvia Serreli and Paolo Calidoni (2017).

Among the experiments spread over recent years, the Junior Summer Schools (JSS)[3] have promoted intense interaction between the subjects (pupils and students, lecturers and parents, artists and volunteers, administrators and citizens) and have offered all participants the chance to experiment with the complexity of real courses of collective learning based on the project themes (Serreli, 2017):

- the educational spaces of the city, in the first JSS edition of 2014 "The city as a classroom";
- new spaces for learning, in the second edition of 2015 "The school I would like";
- the intermediate multi-ethnic spaces, in the third edition of 2016 "Let's play with the world".

In the first edition in a sequence of workshops and short teacher-led lessons the young participants were involved in designing ideal spaces for children, which included in particular open spaces and spaces for public use. With the aim of pinpointing the educational environments of the city that are part of the cognitive world of the youngest inhabitants, the workshops investigated the concept of figurability and fruition (Clemente, 1974), seeking to make places emerge where the children themselves could feel free to produce spaces to which value should be attributed, both by individual original development and by group creativity. In a nutshell, from the words of the children and youngsters motivation and gratification arose in experimenting new creative possibilities for the habitable spaces, and at the same time a great need for spaces of freedom, not just physical. This suggests a lack of facilities in the city and points out a feeling of affliction due to the rules and limits imposed by urban safety (Forni, 2002; Tebaldi, 2016). The "city as a classroom" aimed to represent the educational environments of the city possibly communicating with each other, beginning with the children, for, as the sociologist Elisabetta Forni maintains, they are "the most sensitive indicator to measure urbanity in the city", but also because speaking of the city "is the best way to speak of children". This correlation upon which the JSS projects were developed followed the perspective of "making room for children" even if "the nature of the informal rituals of sociality and social control, the street, square and park have come under discussion more and more over recent decades" (Consonni, 2007).

"The school I would like" concentrated on designing school spaces and the need for a close relationship between learning models and space organisation. The comparison between educational spaces and methods stimulated critical reflection on the variety of methods in different time/space learning contexts: educational spaces and models of the past (in the reading proposed by the Third Age University students), current models linked with the new technologies (Calidoni & Ghiaccio, 2015), and educational specificities coming from different cultures (in the stories of some migrant citizens with regard to African schools). In this respect, in the children's drawings their projects for the habitable spaces of the school confirmed, on the one hand, the rooting of consolidated models and the difficulty of imagining alternative solutions to the traditional classroom, which emphasises the limits of solely *teacher-oriented* approaches (Lisimberti & Montalbetti, 2016). In other solutions the children showed original independent thought and proposed, as various authors have described, an overturning of the

---

3. The UBU activities developed in Alghero, Sassari and some other municipalities in northern Sardinia took place on four occasions during the academic and scholastic year enhancing some of the events of the city: the World Convention on the Rights of the Child in November, World Water Day in March, World Play Day in May and the Junior Summer School in July. The summer schools in particular provided educational, formative and ludic activities organised by the Department of Architecture, Design & Urban Planning. They addressed children aged 7 to 10 years, who took part in three days of lessons, games and design labs at the University facilities and in the public spaces of the town of Alghero. The results of the different initiatives are illustrated in Serreli S. Calidoni P. 2017.

classroom paradigm with new configurations of spaces where different learning modes could take place individually and collectively, involving the experiential, emotional and cognitive sphere of each child. But the space on which there was unanimous agreement was external space, a novel context for learning in the current organisational logics of many schools, which no child forgot to draw.
"Let's play with the world", the third edition of the summer school, was devoted to educational activities reflecting the building up of new urban neighbourliness. The title was chosen by the children and summarised the intercultural experiences the university and schools shared during the academic and scholastic year. The third edition actually applied *student-centred learning* principles, with innovations in teaching compared with the previous editions and favouring moments of freedom for children, so as to render them "more co-responsible for the progress of the process and its outcomes than in traditional didactics" (Lisimberti & Montalbetti, 2016). Designed almost solely by the participants of the previous editions, the third JSS edition proposed among its various activities an interpretation of the territories of the different continents in a maxi-puzzle suggested by the children, a representation of unusual elements that they themselves chose together with teachers and various migrant citizens[4], to talk about and learn of places not yet involved in the teaching syllabuses of the primary school and actually experience the intercultural dimension (Mantovani, 2004). The search for intersecting spaces between the different cultures as the objective of all the project workshops was favoured by the active, creative participation of the young migrants accommodated at the Initial Reception Centre in Alghero, whose placing at schools and university would not have been able to be accompanied by the institutionalised welcoming methods. In this intermediate world represented by the JSS all this constituted a privilege for the children, even before the adults: direct, spontaneous knowledge of the great issue of migration concerning first of all people sometimes a little younger than themselves. For the migrant citizens the JSS was an opportunity to experiment on themselves and their capacity to learn and create a comparison in cultural environments often very remote from the contexts in which they were born.
The summer schools can be configured, quoting De Carlo, as "significant spatial events" that nurture the curiosity of children and adults who enquire about the crucial problems of the world, going beyond knowledge that is fragmented and in compartments, to reflect on the principles of relevant knowledge (Morin 1999), which interprets reality having the prospect of a better future ahead.

Which educational spaces for an intercultural society

The Children's University has been successful from the moment it was conceived as a "space in which ideas like 'public good', 'fair society' or 'shared values' can originate and take shape" (Bauman, 1999), and as an educational environment that welcomes cultural diversity and the plurality of individuals (Morin, 1999). These themes are explored through intercultural educational activities which, by acting on the inter-subjective sphere of the actors, enable other subjects to be intercepted apart from schools and the university (artists, economic operators, social cooperatives, other institutions, cultural associations, citizens, etc.) who would otherwise be excluded from the traditional educational flow.

4. Young migrant citizens at the Alghero Initial Reception Centre who participated all year in the AFRIKASIALGHERO group cultural activities.

The experiments illustrated foster new skills to interpret the city's spatial and social contexts, redefining the role of the university and scholastic institutions in the innovation of educational processes, and singling out and furthering research programmes tending towards the development of design approaches that are both pragmatic and interpret the increasingly transversal, multidimensional and transnational dynamics of the city.

There is a need to be equipped with the appropriate knowledge to interpret territories more characterised by movement than stability, and to encourage action also in environments where uncertainty and unpredictability are constituent factors of reality. The experiments UBU proposes thus undertake to make educational requirements emerge that respond to the current demands of society, even though not previously investigated by the institutions responsible for learning (Comprehensive Scholastic Institutes, University, etc.). The needs are revealed in the interaction of the different subjects taking part in the actions proposed, in sharing an idea of habitable city that may be reinvented by fulfilling hopes (Appadurai, 2011) and reconsidering urban environments as spaces of contact (Choay, 2003) that modify behaviour and lifestyles. Hence, the UBU activities are precious instruments of learning and personal and collective education, also from the multi-ethnic point of view, a quality that cities, especially the smallest ones, build up with great commitment and effort[5].

The UBU cultural project therefore attempts to encourage the development of an open society that learns from the experience produced on each occasion by a set of educational events whose extension is greater than that produced by the single actors who enter into play: "the project, the practices, are not the fruit of dynamics "devoid of a subject" or that can be traced back to some "collective" actor difficult to define. They are always the outcome (not entirely configurable) of actions put in place by intentional ("strategic") actors who may (or may not) agree on objectives and "views" (Pasqui, 2005). Thus the continuity of the events proposed by the Children's University has attempted in these years to favour new perspectives for the "educational" dimension of the individual and social life of the territory we inhabit.

5. To respond to the challenges emerging regarding *living* generated by the arrival of migrants in the town and the appearance of Temporary Residence Centres (CTA) in Sardinia, some Sassari University DADU researchers in Alghero began an experiment on the possibilities of urban planning and architecture action on fluctuating territories. In the research project "Territori e culture in movimento" ("Territories and cultures on the move") promoted by the DADU_UNISS LEAP Lab some theatre companies and various cultural associations are taking an active part in the process towards an intercultural dimension with the involvement of young migrants. See http://www.architettura.uniss.it/ita/la-Scuola/Notizie-e-segnalazioni/Protocollo-di-Intesa-sulla-Multiculturalita

References

- Appadurai A. (2011), *Le aspirazioni nutrono la democrazia*, Et al. Edizioni, Milano.
- Bauman Z. (1999), *La società dell'incertezza*, Il mulino, Bologna.
- Calidoni P., Ghiaccio M.F. (2015), *Viste da vicino, Dinamiche e criticità dell'innovazione digitale nella didattica*, Pensa MultiMedia Editore, Lecce.
- Choay F. (2003), *Espacements: Figure di spazi urbani nel tempo*, Skira, Milano.
- Clemente F. (ed) (1974), *I contenuti formativi della città ambientale*, Pacini, Pisa.
- Consonni G. (2007), "I bambini e la città". Online http://archivio.eddyburg.it/article/articleview/10226/0/29/
- Dettori F. (2017), "Ambienti scolastici e approcci dello *student voice*", in Serreli S, Calidoni P. (eds), "Città e formazione. Esperienze tra Urbanistica e Didattica", FrancoAngeli, Milano
- Fabbri L., Melacarne C. (2015), *Apprendere a scuola. Metodologie attive di sviluppo e dispositivi riflessivi*, FrancoAngeli, Milano.
- Forni E. (2002), *La città di Batman. Bambini, conflitti, sicurezza urbana"*, Bollati Boringhieri, Torino.
- Kember D., Kwan K.P. (2000), "Lecturers' Approaches to Teaching and their Relationship to Conceptions of Good Teaching". In Hativa N., Goodyear P. (eds) *Teacher Thinking, Beliefs and Knowledge in Higher Education*, Kluver, Dordrecht.
- Kember D. (2008), "Promoting Student-Centred Forms of Learning across an Entire University, Higher Education, 58 (1) p.1-13
- Lisimberti C., Montalbetti K.(2016), "Con le mani in pasta. Apprendere la ricerca fra teoria e pratica", in Luzzato G., Turri M. (ediitors), *Innovazioni didattiche nelle riforme universitarie*. Scuola Democratica "Learning for Democracy", Special Issue 3/2016 p 635-644, Il Mulino, Bologna.
- Mantovani G. (2004), *Intercultura. È possibile evitare le guerre culturali?*, Il Mulino, Bologna.
- Morin E. (2001), *I sette saperi necessari all'educazione del futuro*, Raffaello Cortina Editore, Milano.
- Morin E. (2015), Imparare a vivere, Manifesto per cambiare l'educazione, Raffaello Cortina Editore, Milano.
- Pasqui G. (2005), *Progetto, governo, società. Ripensare le politiche territoriali*, FrancoAngeli, Milano.
- Serreli S, Calidoni P. (2017), eds "Città e formazione. Esperienze tra Urbanistica e Didattica", FrancoAngeli, Milano
- Serreli S. (2017), "Università dei Bambini: interpretazioni e percorsi", in Serreli S, Calidoni P. (eds) "Città e formazione. Esperienze tra Urbanistica e Didattica", FrancoAngeli, Milano
- Sisti F., Torrisi G. (2016), "Il puzzle dell'innovazione didattica all'Università di Urbino", in Luzzato G., Turri M. (editors), *Innovazioni didattiche nelle riforme universitarie*. Scuola Democratica "Learning for Democracy", Special Issue 3/2016 p.687-705, Il Mulino, Bologna.
- Tebaldi M. (2016), *Sicurezza, libertà e democrazia*, il Mulino, Bologna.
- Torrisi G. (2013), "Augmented Learning Uncovered: Five E-learning Solutions Tested and Compared", in ICERI 2013 Proceedings, Servilla, ICIERI Publications.

UBU_Junior Summer School (2016); Let's play with the world, Alghero, DADU

3.4

Spaces as learning devices

Beate Weyland*

The way we perceive school space has changed in recent years, evolving from a standard place of teaching to one where people live and learn, and therefore, inhabit. Built by designers according to the client's wishes, school communities have adapted without paying great attention to the quality of the spaces. Schools have become the subject of architectural and pedagogical commitments, the responsibility of which lies with all entities involved in schools: administrative bodies, school leaders, teachers, parents and citizens.

Linking pedagogy to architecture

Projects involving close dialogue between pedagogy and architecture (Watschinger - Weyland, 2017; Weyland-Attia, 2015; Ceppi, Zinni, 1998; Montag Stiftungen, 2012; Lippman, 2010; Hille, 2011; Schratz 2006 to mention a few) investigate specific influences of the environment on learning, starting from the specific needs of learners. Movement – the ability and freedom to explore the school, utilising all spaces to learn and discover, is becoming increasingly valued. Contemplation – the necessity to allow private spaces where individuals can retreat in order to study or think. Sharing and cooperation – following the concept of school as a community, some school environments (empty and anonymous entrances, wide and barren corridors) are being re-envisaged as spaces for groups to gather. Attention is given to cooperative learning and aspects such as collaboration, responsibility, solidarity and the contribution of the single student to the group, are enhanced. Identity – the classroom is the identify space of students of a class. It is the starting point from which to explore the whole learning space. It can also be considered the place where input is provided, discipline maintained, or specific groups find their identify. School is confirmed as the most appropriate platform to rethink the learning process. Traditional teaching is in crisis, as schools are confronted with times, methods and content of increasing diversity. To evolve towards more open and plural methods of teaching, it is essential to develop new didactic approaches and to modify actions and spaces in tandem.

Dwelling schools

Several studies (Seydel, 2016; Gray, 2015) indicate both pleasure and motivation as key elements that stimulate learning. Playing, in this sense, seems to be the macro category to rethink both teaching and learning approaches and spaces. If the teacher-student-content relationship is raised in a playful form and promote an active involvement of all stakeholders, also the environment needs to be rethinked. How to create comfort conditions, informal and recreational environments, that stimulate learning and otherwise complies with the institutional frameworks of public schools?

* **Beate Weyland, Free University of Bozen-Bolzano.**

Schools that wants to be "dwelt" and bridge from traditional to innovative solutions, should accommodate items, tools, and content that have not been studied for her, but that belong to other contexts. Innovation in architecture and in educational furnishing (mandatory to meet new educational needs) goes through the overcoming of the traditional model (school needs specially-crafted furniture) and needs to be guided with a pedagogical concept.

There are in Italy three key variables liked to institutional documents and laws to be considered in the processes of accompanying schools to change their spaces: the *organizational variable* (schedule, teachers, external collaborations) regulated by the law 59 of the 15th of march 1997 of school autonomy and the law 107 "The Good School"of 2015; the *variable of contents and programs*, which refers to the national guidelines for the nursery, primary and secondary school of 2012 (http://www.indicazioninazionali.it); the *infrastructure variable* (spaces and furnishings) which has as a reference point in the National Law of 1975 on school building and in the Indications of the school building of 2013 (www.ediliziascolastica.it). It is necessary to act on all three variables to develop new models. One depends from one another. One activates another one. But it is also important to expand the framework within which a school is designed. Planning a space means giving an answer for specific functions but above all it is establishing relations between systems of meaning. Planning between pedagogy and architecture begins as a "relational art", a dynamic between local cultures and instances of existence, including the identity of a society and the environment that it inhabits. This sort of approach has elements of rationality, desire, emotions, memories, creativity, and links the element of a space with our experiences in a complex web of relations between scales of intervention, actors involved in the process and issues to address. Space is also seen as a sensor of new forms of territoriality and lifestyles. It is the basis for a fruitful interdisciplinary exchange in which planning becomes the matrix for communication and cooperation between the different scales that operate on the school.

The school is the most appropriate laboratory for learning about the capacity of interactive citizenship and educational cooperation. In order to respond to the changes in more recent generations of students – in their values, their relationship with knowledge and culture – the different actors in the educational process can find, in the shared planning of the teaching and learning profile of the school, an ideal way to make their work more effective and efficient. When they participate in the building of a new school or the re-structuring of an existing one, the teachers and head teachers, the families and the educational community understood more broadly, can help redefine what the school will be and what they are willing to do to reach their educational, didactic and cultural objectives.

Nationally and internationally we find that educational communities are able to define their own identity and define their own way of doing things (that is, the teacher, student, knowledge-culture relationship) can also orient planning choices and give birth to new models, especially with regard to the organisation of internal spaces (Marcarini, 2016). A notable example comes from schools in Reggio Emilia, which, based on the "Reggio approach", always include in the building's plan a large central space for joint activities, a large workshop for manual and artistic work and smaller workshop spaces in the different sections so as to guarantee the continuity of didactic-educational work with the child's expressive-creative dimension (Weyland, 2014). This is distinguished by its attempt at transparency and presentation of a visual space at children's height so as to respect the pedagogical principle of a working community. The concept of "school without a backpack" in Tuscany (www.senzazaino.it) highlights the importance of creating the classroom as a lived space, where not only the course material always remain at the students' and teachers' disposal, but also all the students' equipment: carrying cases, folders, books, notebooks, brushes and compasses. These choices result in a reconfiguration of the classroom space and furnishings across two inter-connected rooms organised along thematic areas and differentiated work corners. These are just two examples, in addition to the classics of the Montessori and Steiner models, where space is defined by educational objectives and in which it becomes a real and proper device to transmit the school's culture.

The school as a whole

A school has the natural inclination to organically group structural entities into complex wholes, which are equipped with qualitatively new properties in comparison to their single components and which boast a higher degree of interaction and complexity within them. A school's parts (from the individual to the group, from teachers to directors, from the child to the families, from education to the territory, from the desk to the classroom or the whole building) can be almost said not to be real, but rather the result of abstract analytical distinctions that never adequately express the concept of "school" — which is something that goes well beyond the mere sum of all its components. The quality of the system of parts is what makes the difference. The complex of all its parts gives a school its identity: the architecture that shapes it, the objects that furnish it, the people who attend it, the teachers who animate it, the directors who manage it, the parents who support it, etc. The educational aspect of a school consists of a meaningful synthesis not just for the pupil but for the whole social community. It emanates the value system of the society of which it is a part and takes steps to become a large center of socio-cultural aggregation — a new agora (Attia, Weyland, 2013).

"A school stands out as a whole or a complexity unitarily woven and intertwined with elements, contributions, facets. Until we have to deal with the event called education, it is impossible to remove one instance of the unitary whole from the elements that make up a school (...). The sum of influences, actions, interventions and operations with which a school expresses its educational presence tends to harmonically transform into a meaningful synthesis for the pupil." (Scurati, 1997, p. 184)

References

- Attia S., Weyland B. (ed.), *Costruire pedagogie.* Turris Babel, Architecture Foundation Alto Adige, nr. 93, 10/2013.
- Ceppi G. Zini M. (1998), Bambini, Spazi, Relazioni, Reggio Children Editore, Reggio Emilia.
- Grey P. (2015), *Lasciateli giocare.* Torino: Einaudi.
- Hille R.T. (2011), *Modern Schools: A centrury of design for education.* New Jersey: Wiley,.
- Marcarini M. (2016), *Pedarchitettura.* Roma: Studium.
- Lippman, P. (2010), *Evidence-Based Design of Elementary and Secondary Schools: A Responsive Approach to Creating Learning Environments.* New Jersey: Wiley
- Noack M. (1996), *Der Schulraum als Pädagogikum.* Weinheim: Dt. StudienVerlag
- Opp G., Brosch A. Ed. (2010), *Lebensraum Schule: Raumkonzepte planen, gestalten, entwickeln.* Stuttgart:Fraunhofer IRB Verl
- Schratz M. (2000), Qualitätsentwicklung. Frakfurt: Belz
- Scurati C. (1997), Pedagogia della scuola, Brescia: La Scuola
- Seydel O. (2012), *Pädagogische Perspektiven für den Schulbau - 12 ThesenÐ - Symposium für einen ÐleistungsfähigenÐ.* Berlin: Tosi L., Borri G. (2019): *Fare didattica in spazi flessibili. Progettare, organizzare e utilizzare gli ambienti di apprendimento a scuola.* Firenze: GiuntiScuola
- Weyland B., Attia S. (2015), *Progettare scuole tra pedagogia e architettura.* Milano: Guerini
- Welling, S., Breiter, A., Schulz, A. H. (2015), Mediatisierte Organisationswelten in Schulen, Springer, Heidelberg.
- Woolner, P. (2015), *School design together.* New York: Routledge

Experiences of research and project:

The school: a pedagogy of buildings

Massimo Ferrari*

Architectural themes can be studied and experienced from different points of view. This possibility requires, first of all, the choice of a scope for our research. It dictates a reflection on the stance that we deem more important and useful to learn more about a topic and to to describe it. Eventually, it also implies the possibility to stick to certain interpretations and to precise critical attitudes. In general, the tools we choose for our research depend directly on our aim, but in most specific cases, it may be the character, depth and clarity of a theme to determine the pace and the study methods, reducing personal interpretations to a mere succession of critical arguments. As architects, we study examples, we search for solutions, we watch and criticize past and present to imagine the future of our branch of knowledge. For this reason, we think that research, critique and projects must necessarily coexist in our work because they take nourishment from one another, and if one of them misses, the other two become sterile. We share the same opinion as those who think that in order to write a novel, you need to read novels, or that you need to study science if you want to understand its rules. They are continually fluctuating between exactness and its representation, and so are we.
Architecture lives in the present and imagines the future.
With their projects - utopian but realistic at the same time - architects are often the only ones who still manage to reinvent the usual ways of life, to question traditions, and to suggest new ways to make a better future. The basis of our work has always been our capacity to be interpreted by the Society that we represent through built shapes. After all, Society is the one asking the questions to which architecture must give an honest answer. This interpretation implies two things: on one hand, a research aimed at acquiring an in-depth and increasing knowledge of the theme discussed; on the other hand, the ideal possibility to imagine places that suit people and their lives, even before thinking about standard compliance and technological efficiency.
In a letter describing his impressions of the Cologne Cathedral, Hegel wrote: *"In such a cathedral there is room for an entire community. For here the whole community of a city and its neighbourhood is to assemble not round the building but inside it.*
For this reason all the various interests of life which touch religion in any way have their place alongside one another. The wide space is not divided and narrowed by series of rows of pews; everyone goes and comes unhindered, hires or takes a chair for his present use, kneels down, offers his prayer, and departs again [...]"

* Massimo Ferrari, Politecnico di Milano.

The tight bond between architecture and liturgy is clearly represented by some sacred buildings of the 20th century. Therefore, we can use them as an example of how a general principle can be turned into something tangible, the proof of the existence of a relationship between a thesis and its interpretation. The same dialectic method can be applied even to address a completely different theme, such as the educational building.
Like for the Cologne cathedral, the laborious investigation of the generic features that make a building adequate is one of the main issues even with respect to the relationship between architecture and pedagogy, which is at the basis of our latest work on educational buildings. Once again, we are faced with the need to rewrite norms and conventions starting from a virtuous relationship that can prove the causal connection between the research of the features that make an ideal learning environment, and the construction of spaces that comply with such interpretations.
In today's world, education and educational buildings *of any order and grade* are one of the main themes from both the political and civil point of view. From the 20th century to the present day, school architecture has anticipated, followed, and sometimes even chased, social transformations, reforms and educational proposals, as if it were a sort of silent counterpoint to the history of our country. At the same time, many school building projects, from nursery schools to universities, have written important chapters in the history of architectures, both on a national and an international level, by overturning consolidated principles in the name of an idea of teaching that was free from any restriction.
Quoting Aldo Rossi, *classrooms and corridors* are training places where the relationship between collective and individual space - between a shape and the teaching idea behind it - become the backbone of an ongoing dialogue between teaching methods and architecture, which has its starting point the work of some of the most important educators of the 20th century, such as Rosa and Carolina Agazzi, Maria Montessori, Loris Malaguzzi and Mario Lodi.
In this sense, Italy has an original story to tell. The widespread interest in applied research and pedagogical studies provides a rich and useful resource, and the analysis of the composition of buildings runs in parallel with the research on the values of education and on how learning spaces can support them. At the beginning of the last century, Maria Montessori, referring to learning places, wrote: *"Education is a natural process spontaneously carried out by the child, and is acquired not by listening to words but by experiences in the environment."* Several years later, Loris Malaguzzi described the atelier as *"a subversive irruption, an additional complication providing an additional tool. It has allowed rich combinations and creative possibilities among the different languages and nonverbal intelligence of children"*. Eventually, in the mid 70s Mario Lodi, wrote about the need to *"create a community where children feel equals, like brothers and sisters"*. In architecture, the environment, the atelier, and the community represent as many opportunities for research; whether they are physical spaces or figurative forms, school buildings generate from them almost by centrifugal force.
The classroom is the minimal module and probably it best represents the most domestic and authentic character of the educational building, the germ of the change that can occur in the future of learning and teaching. Once again, the example comes from the past: the American one-room school. This model was later exported to other countries including Austria, Germany, Australia and Ireland, and rather than a training goal, it represents a suggestion or a figurative horizon. At the end of the 19th century, these little rural schools were built not far from the town centre; they were surrounded by

nature and consisted of a single room. Everything was reduced to the minimum: one staircase, one entrance door, one classroom, the restrooms. Children of different ages sat in the same room, and the same teacher taught reading, writing, arithmetic, history and geography. These little buildings, with their simple shapes, often became the centre of the entire community.
When we say we should think about the school of the future, this is not a catch phrase, but an invitation to take inspiration from these central examples - quite frequent in the last century - where all disciplines were in perfect sync. And since we believe in the critical potential of dialogue and discussion, we believe in the capacity of a new generation of architects to raise awareness and boost the quality of Italian architecture.

Of every order and degree. The architecture of the school; exhibition conceived and produced by Triennale di Milano, curated by Massimo Ferrari with Claudia Tinazzi, Beatrice Gerli, Daniela Mauer, Como 2015.

Di ogni ordine e grado
La Triennale di Milano e la Scuola
1933
La V Triennale di Milano
Architettura, abitazione,
arti decorative e industriali
Di ogni ordine e grado
L'architettura della Scuola
Confronti

Learning open space

Gianluigi Mondaini*

"Good schooling begins with a beautiful school." Thus Luca Molinari titled one of his two weekly columns in *L'Espresso*, recalling the recent school reform put forward by the Renzi government – "the good school," in fact – but above all the need to put the "beauty" and quality of architectural spaces at the centre of the discussion (which usually feeds on regulations, cuts, numbers, roles, articles and assorted paragraphs and clauses). Spatial experimentation, after years of standardized types and generalized speculation leading – in Italy in particular – to the symbolic irrelevance and collective unrecognizability of school buildings, standing in the post-modern territory of indifference, ought once again to take pride of place in every deliberation, if a country seriously intends to take its own future in hand.

The subject of school buildings is undoubtedly a specialized research field, both with respect to the project itself and to several other matters, such as the country's path of development chosen in the realm of politics. The necessary synergy between the two can and must convey – through the school building, but also through any building that constitutes an expression of the community – its primary meaning, the specificity of being a "community asset," the symbolic locus of society's development. Schools, therefore, pose a dual challenge: first, the relationship between building and context that enables planning of the urban area, that may characterize the public space, and that will have an impact on the people who inhabit that particular area; second, the building's architectural quality, with its new functional and environmental standards that will have a major role in shaping the students' perception of comfort.

Concerning the first challenge, the school building is a reflection of a country's body politic, just as any other public building whose physicality attests to its role as common asset and instrument of shared ethical values. In recent years we have witnessed a degenerative backslide of instructional spaces that closely parallels the social and economic decline (from the free-market infatuation onwards); the government – but also the silent intelligentsia – failed to grasp what the globalization of economic systems and ideas entailed. The confusion between liberalization and weak thought and the speculative separation of form and content led, in typically Italian fashion, to the loss of value (not only economic) of things, spaces, buildings and especially the concept of "community asset". Gradual decay has eaten away at that uniquely Italian evocativeness that some buildings could convey through a formal and emotional relationship with an urban plan that balanced civic representation and physical interpretation of the environment. Alas, this model has been lost. In contrast with other countries, Italy has been singularly incapable of valorising public spaces through the necessary synergy between public and private investment; public buildings have been relegated to an increasingly marginal role, placed "speculatively" in the outskirts, deprived of all community, landscape or urban value.

The challenge we face is to find a way to use the school building as positive energy to regenerate those city areas that must necessarily be seized and conquered, redeploying the historical Italian ability to build city spaces through an active dialogue between building and space; a dialogue once pregnant that is now undoubtedly open and void, yet may still be amenable to design through architecture. An unrenounceable "project" that should, of course, be an opportunity to recycle existing building and spaces, but that will

*** Gianluigi Mondaini, Politecnico delle Marche.**

also have to reinstate quality to its central position: an unrenounceable investment on the appropriateness of things, and on their aesthetic value in particular. Schools are active places where diverse actors engage one another; if the building's architecture succeeds in grasping this peculiarity and transforming it into a space, then the school building can indeed extend its influence to the surrounding context.
The second challenge to be faced by every actor involved (architects, administrators, teachers and citizens) concerns the need to re-invent spaces, enabling them to incorporate meteoric societal changes and, therefore, to cater to the needs of students; these needs will in turn arise from teaching practices and methodologies that should consider the spatial context as an essential element. We have reached a point where rising to this challenge and its associated demands has become mandatory, if we want to rethink the school building from the inside, to update its space in accordance with new demands, to endow the new educational aims with substance and form, and hence to bring about its "civic rebirth". "School is where that miracle happens that turns subjects into citizens" is the well-known maxim of Piero Calamandrei;[1] thus, the aesthetics of the spaces where students spend most of their time must be active, vital and stimulating, and provide through its physical attributes the possibility of experiencing a higher dimension of the things that surround us.
Space can become an educational programme, and in the "agora" – using a metaphor that sets the city's dynamism as a reference point for schools – a miracle may take place whereby children and students learn, in addition to the theories behind phenomena, the practice of civilness, which will be positive to the extent that the physical experience is pleasurable, conveyed and stimulated as it is through a beautiful, captivating space.
Implicit in the concept of education is the desire to form, to give shape; not only in the physical sense, to be sure, but also as an expression of the intimate relationship between education, shape and space. And it is precisely because the city and society have in time undergone profound changes that the spaces of education must in turn change, in order to denote the complexity of such transformations and determine their new form. We need to recapture the experimental dimension that has characterized the best of our architecture, where it was precisely the construction of school facilities that constituted one of the most innovative sectors, both in terms of architecture and educational principles. For all that our schools were once at the forefront of innovation, the current school building panorama in Italy is one of the direst in Europe, having lost during the second half of the 20th century all the foundational originality of its inception.
Along with the relationship between school and urban context, it is necessary to reflect upon a veritable revolution that forces us to pay due attention to the school building's internal spatial configuration. With the stimulus of innovative teaching methods, that revolution finds in pedagogical action an ally for reconfiguring the typological codification of the school building. The matter at hand is multifarious, ranging from a radical rethinking of frontal instruction to the new role of open spaces and connectivity, of services and green areas, issues that were once subordinate to the central role of the classroom. This enlarged field greatly extends the range of architectural possibilities, and stimulates opportunities for a renewed, more open dynamics of education.
Pedagogists and the architects that interpret their theories and translate them into forms, are finally starting to imagine more complex and more stimulating spaces and scenarios, a sort of "scattered learning".

---

[1]. Piero Calamandrei, (1889-1956): Italian jurist and author.

We have to imagine new, less delineated, more complex and more stimulating spaces, designed for a different kind of concentration and interaction. We have to go beyond the "typical" school building, so entrenched in each of us and synthesized in the relationship between corridors (interaction) and classrooms (isolation).Today the need emerges for a different perspective that sees the school as a single unified space, resulting from the superposition of several environmental matrices: information, relations, spaces, architectural components and materials. In addition to the various specific functions, we also have to define the modes of relationship between them; if the new school is to become "an environmental matrix for learning," we propose a fluid connexion between its constitutive elements. This dimension of fluidity finds in open spaces such as the horizontal and vertical connexions and the atrium the "new square," the new setting where this redefinition becomes tangible. The square in particular may the hub of all of the school's public functions, and become the locus of informal interactions; in the case of larger schools, the square may become a veritable "agora," the symbolic location of dialogue and exchange that sustains the development of knowledge within the school-city. The fluid and dynamic dimension of the learning space may be further advanced by a new quest for transparency, in both the figurative and concrete senses. The building's spatial and material transparency provides a paradigmatic backdrop to multiple themes, all of them capable of investing the contemporary school building with a new significance.
Open spaces, and green areas in particular, are one of such themes; when judiciously deployed outside and inside the structure, a fruitful interplay of solidity and lightness can be obtained, further enhanced by contemporary visions of artificiality and naturalness that transparency allows. Green areas may rightfully constitute project material, from the point of view of pedagogical action and of their pertinence as a passive strategy to optimize comfort and energy savings. Indeed, along with the relationship between artificial spaces and the strategic use of nature, the new project paradigm will be the building's sustainability. And we are not referring to the performance of technological solutions, but to take action on the project itself with common sense and good ideas, optimizing form and quality and eschewing the marketing overtones so often associated with the word "green" when applied to the word "building". The required cultural leap is already under way, using "site specific" spatial and architectural strategies instead of the global models that characterized the times of myopic energy waste.
An aspect that deserves particular attention when designing educational spaces is the shift from their heroic dimension (in the sense of an ideological and disciplinary superstructure imposed upon the specific qualities of the students'' vital space) to the ludic, alluring dimension these buildings should have in order to establish an immediate, positive and empathetic relationship with their young users. Light, colour, diverse morphologies and multiplicity of materials were set aside form modernity onwards and almost wholly excluded from the rulebook, with the exception of the disastrous post-modern period.
Colour, free form and appealing sequences, when not a mere restatement of stylistic codes, may constitute positive strategies to reach a state of harmony both with the context and with the students the building is meant for. Colour helps us inhabit a space in a more intuitive fashion and keep away from the monochromatic modernist rationality, fostering a more joyful, more dynamic spatial configuration, in keeping with the complexity of our contemporary environment. Colour, graphics, texts and images often mediate between users and spaces, rendering the space more approachable and hence welcoming and familiar. The perception of space and the modalities of its lived experience are particularly important factors for sustaining the relevance of an educational project that can engage students emotionally, fuelling their imagination and fostering a pleasurable physical well-being. A feeling conveyed through the multiplicity

of formal suggestions, the configuration of spatial sequences and the material and chromatic peculiarities that stimulate curiosity and knowledge; these are architectural strategies consistent with the cognitive and ludic rules expressed through the seductive techniques of play.
Educational spaces must be enhanced by shapes and spaces that, acting in consonance with an engaging methodology, communicate the experience and integrate it into a narrative, as in a tale that recalls the adventure of life and its future and becomes, through the required sensorial stimuli, something physical that can be experienced and perceived through all the senses.
Ricardo Carvalho, commenting a special school building plan put forward by the Portuguese government, emphasizes the strategic need for a school project to have "as its aim the creation of a humanized, optimistic environment, endowed with a lyrical and poetic dimension and capable of forming a community's collective memory".
Many of the considerations put forward in the preceding discussion constitute the basis of the architectural concepts underlying forms and spaces that informed the conference held at the Arghero Summer School ILS for two schools built by Mondaini Roscani Architetti Associati. The first is the result of a bid for a school complex at Montecarotto, Ancona, of which the school was to be the first stage; the second is a nursery school for the Consorzio per lo Sviluppo Industriale di Monfalcone, Staranzano, Gorizia; this too was the first stage of a tender for a service hub in that industrial area.
In the case of the first school, the central theme is the building's striving for anchorage to the surrounding context and landscape, and at the same time the ability to stimulate the children's imagination through a multiplicity of unusual forms, cavities and colours that recall a compelling ludic dimension. The terrain's steepness is put to good use by a suspended walkway and a colourful concrete entrance block which constitutes the architectural extension of the bridge. Ramp, gangway and entrance block constitute, along with the stairs, the perpendicular axis that projects the heart of the building towards the valley below, eliciting a physical perception of belonging to the territory, its specific features and environmental qualities. This two-storey heart characterizes a multi-function spatial configuration, from reception to vertical and horizontal passageways that also become terraces for small collective events. All distribution spaces have been designed as open and generous places for meeting, playing and sharing, thanks also to the cavities of the leaning wall on the main front which irradiate the space with a pleasant play of lights of varying size and colour.
The Staranzano nursery school project combines two themes: the relationship between space and education, and the search for sustainability. The building envisages a typological re-invention of the space dedicated to children, releasing it from its physical, material boundaries, opening up the schoolroom space and immersing the outside on the inside; thus, a dynamic and stimulating educational unit emerges where the external and internal spaces merge in the continuous physical and visual border between artefact and nature.
A series of patios, visually connected to each other and to the external garden, conceived as an open air extension of the play area, allow a typological re-invention that enriches the space, making it more stimulating and available for ludic-didactic creations. In addition, the shell's shape provides, of course, the correct general orientation of the building, but also turns and folds, and through a surface that twists in two directions becomes, by shape and inclination, not a sterile support for technology, but itself a passive strategy to respond to the wind, to the sun, and characterizes the building's architectural outline.

PRIMARY SCHOOL "G. MAGAGNINI"

Montecarotto, Ancona - 2009/2014

Author: Mondaini Roscani Architetti Associati - Gianluigi Mondaini, Gian Paolo Roscani
Works Management: Mondaini Roscani Associate Architects - Gian Paolo Roscani
Collaborators: Silvia Santini, Mauro Sorrisi, Annalisa Buscarini, Claudia Ciavattini, Christian Rubini, Fabrizio Belluzzi
Consultants: Structures project: Ing. Roberto Morici
Technologic system: Sigea Srl - Ing. Elio Ottaviani, P.I. Alessandro Franchini
Client: Municipality of Montecarotto, Ancona
Construction company: MEG Costruzioni Srl, L'Aquila
Photographs: Marco Capannini, Jesi-Pesaro

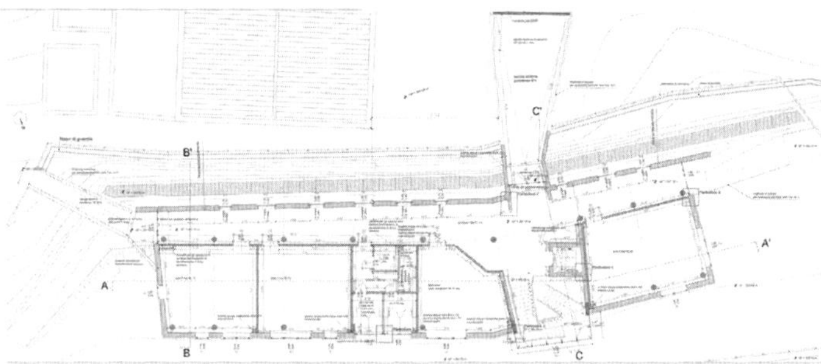

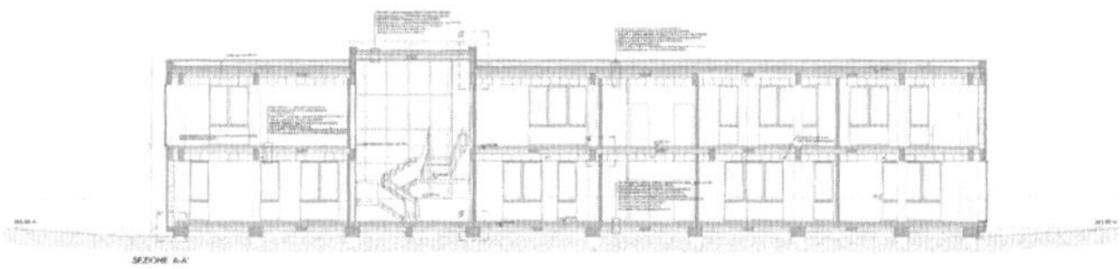

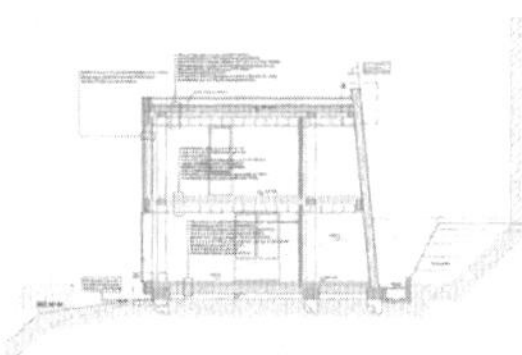

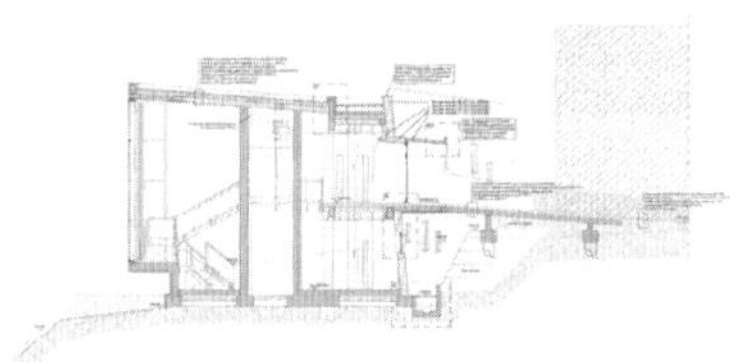

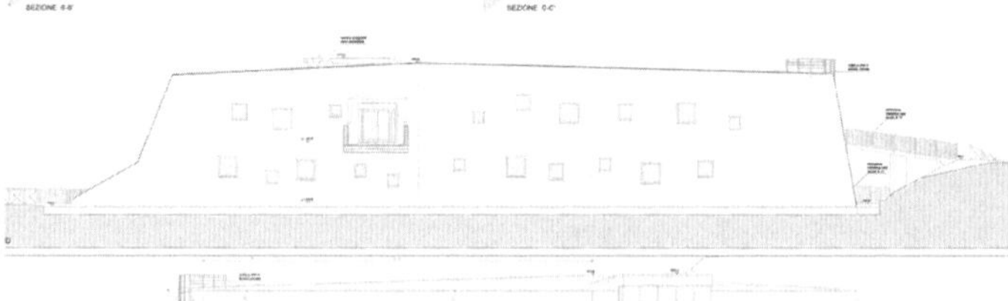

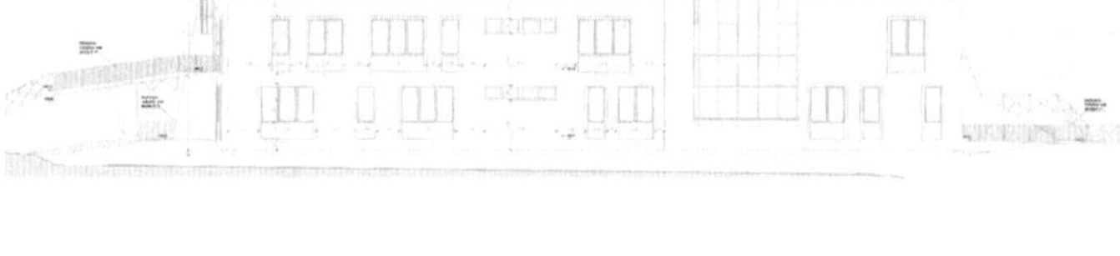

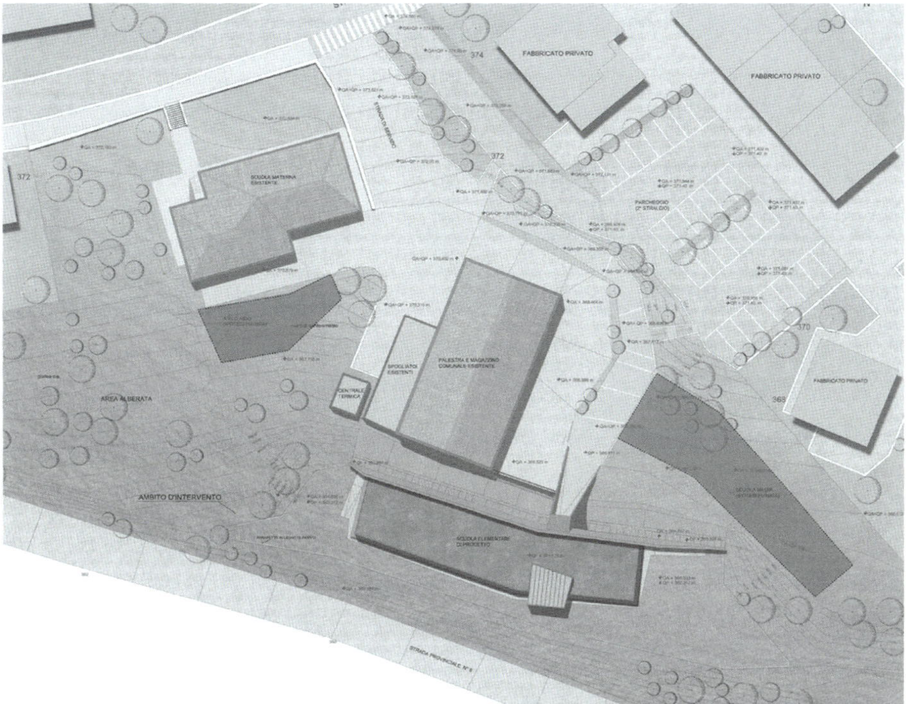

General plan and north elevation of the school

The school in the hills

Entrance hall to the school

Interior spaces and distribution

KINDERGARTEN

Staranzano, Gorizia - 2010/2016

Author: Mondaini Roscani Associated Architects - Gianluigi Mondaini, Gian Paolo Roscani
Works Management: Mondaini Roscani Associated Architects - Gian Paolo Roscani
Collaborators: Silvia Santini, Mauro Sorrisi, Annalisa Buscarini, Claudia Ciavattini, Christian Rubini, Fabrizio Belluzzi
Consultants: Structural design: Ing. Roberto Morici
Plant design: Sigea Srl - Ing. Elio Ottaviani, P.I. Alessandro Franchini
Client: Municipality of Montecarotto, Ancona
Construction company: MEG Costruzioni Srl, L'Aquila
Photographs: Marco Capannini, Jesi-Pesaro

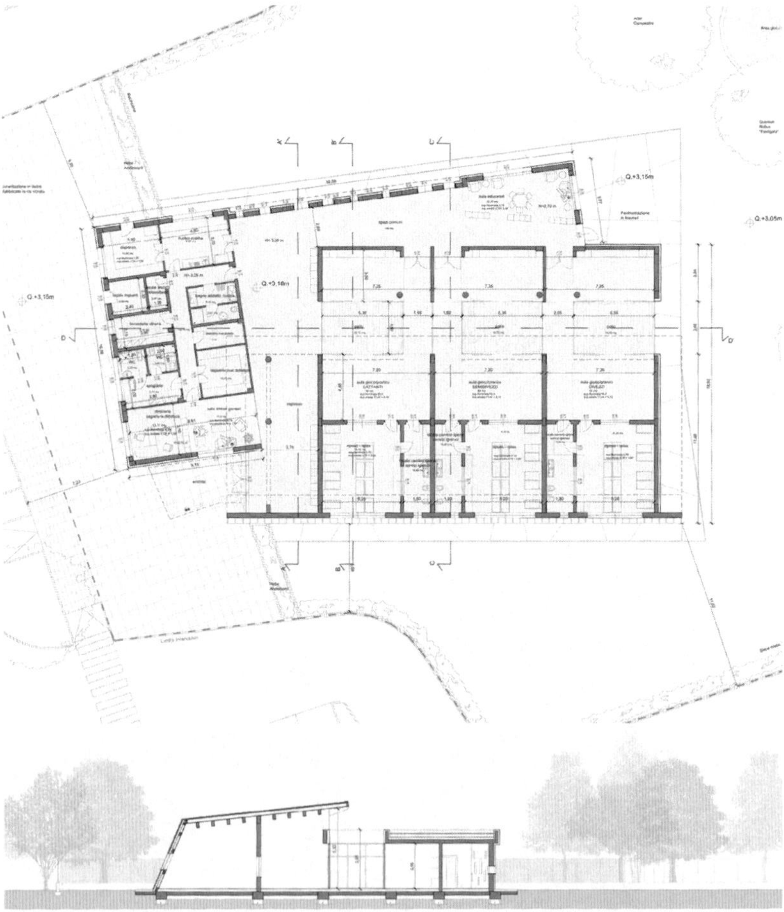

Plan and section

General plan and elevations

The garden

The entrance space

The patio as an extension of the learning spaces

4
A school for each and everyone

@UP School

Recent discoveries in the field of neuroscience have shown a relationship between motor patterns and higher cognitive functions; an action is no longer understood as the result of a phase of perception and subsequent interpretation, but as an integral part of a process in which all the components cooperate simultaneously. Space provides affordances – reasons for action – and the subject, through their cognitive system, takes opportunities based on a kind of wager, a forecast, as to how that space will be in the future. It is the action that leads to knowledge, and space supports this process in a decisive way.

#opportunities for action #cognitive functions #space #dynamic form #learning tests

Focus Group the school for everyone and for everyone.

Spazio
Educazione
Tecnologie

4.1

The neurodevelopmental disorders: knowing to integrate

Alessandro Zuddas e Silvia Petza*

Over the last few decades, classical learning spaces have been significantly re-defined, moving from uniform spaces in which the furniture was often the only variation between the different contexts, to flexible articulations of space defined by modular and multifunctional environments in the attempt to adapt different pupil's subjective experiences in the acquisition of knowledge. Parallel to such transformation of learning spaces new teaching approaches have been developed: currently rather than a uniform teacher –pupil knowledge transmission, a direct approach to knowledge (i.e. every pupil must discover through direct experimentation) is preferred; these new defined approaches to knowledge implementation may, at least in part, attenuate specific differences in basic knowledge related to the social environment of origin.

Learning and severe neurodevelopmental disorders

Although the principles of this conceptual evolution have become a common background for teachers and planners who interact with students showing a "typical" neuro-cognitive (and socio-affective) development, many issues still remain unresolved when users of innovative learning spaces include subjects with neuro-development disorders.

The modes of interaction with people (peers and adults of reference) and space by pupils with two specific neuro-developmental disorders may be discussed as an example: autism spectrum disorders (ASD) and deficit disorder with hyperactivity (ADHD).

Autistic spectrum disorders are clinically characterized by impairment in the development of communication and social relationships, and by the presence of restricted and repetitive interests and activities. The main, though not exclusive, neuropsychological alterations are represented by a deficit (on a neurobiological basis) in mental ability to perceive that others may have different thoughts and feelings (so called "*therory of mind*") associated to a deficit in the *internal coherence* (i.e. the inability to merge the individual components of an image to define its overall meaning/ value).

These deficits in *theory of mind* and *internal coherence* , can cause difficulties in abstraction and in reasoning by abstract concepts, which in turn induce difficulties in predicting the behaviour of others (often perceived as "*objects*" rather than "*people*") and consequently a state of tension, anxiety and concern in unfamiliar and difficult to foresee environments. Moreover, narrow and repetitive interests may also include specific abnormal sensory perceptions such as hypersensitivity to the sounds,

* Alessandro Zuddas and Silvia Petza, Clinic of Child and Adolescent Neuropsychiatry University of Studies of Cagliari, Department of Biomedical Sciences, & Pediatric Hospital "A. Cao "; Hospital Company "G. Brotzu ", Cagliari.

idiosyncrasy reactions to specific colours, particular interest in the presentation of light stimuli (eg "blades of light"), abnormal interest or aversion for tactile or olfactory stimuli.
ADHD is characterized by high levels of attention, hyperactivity and impulsivity compared to peers: the main neuropsychological alterations are represented by impairment in selecting appropriate environmental stimuli ignoring background and inappropriate / distracting stimuli, in inhibiting the automatic responses; different time and reward perception is also present, frequently leading to *delay aversion*.
ADHD clinical manifestations change with age and developmental level; the disorder is frequently comorbid with other psychiatric disorders (such as oppositional defiant behviour , conduct or mood disorders), and it predisposes to further psychopathology in later life such as depression, substance use disorder, antisocial personality). Unlike other developmental disorders, for ADHD international clincal guidelines have defined as highly effective therapeutic strategy to combine pharmacological intervention with specific psychoeducational approaches
Recent estimates indicate the presence of Autism spectrum disorders with variable severity in at least one individual over 100; a variable severity of ADHD has been reported in at least 2 to 4 individual over 100. For both disorders, clinical manifestation may also vary according to gender, age and other variables such as the cognitive level.

Space for learning in children and adolescents with neurodevelopmental disorders

In children with ASD, deficits in the *theory of mind* and in *internal coherence* may explain the difficulties not only in social communication, but also in perceiving and defining the space and the time: a predictable environment becomes therefore essential.
On the other hand, for a child with ADHD the difficulties in selecting the environmental stimuli and in inhibiting automatic responses along with the different perception of time (which is accelerated compared to peers and adults, leading to "boredom" if not enough stimuli are available) make essential for an adequate learning process the rapid but modulated presentation of salient environmental stimuli and their rewards, with a decrease in time devoted to the "explanation" and an increased total time dedicated to planning experience (albeit subdivided into short and repeated time units).
Applying these considerations to the definition of innovative learning spaces is therefore a major challenge for clinicians, teachers and architects.
Possible innovative solutions may include the modulation of the classroom space by adding smaller environments adjoining to the main room; spaces may also be differentiated by size, colour or furnishing or any visual signs that help the child to safely and predictably define the activities to be performed; such spaces should also be able to allow temporary customized teaching methods (also in 1: 1 or 1: 2 teacher-pupil ratio).
As mentioned above, the experience of "*discovering together*" is essential in modern (psycho) educational strategies: in children and adolescents with neurodevelopmental disorder is therefore crucial to define both "curricular" and "ludic" learning spaces in order to allow both group sharing of the experiences and the individual internalization of the sames.
To make easier their experience, the general spatial organization of the entire school also plays a fundamental role: to improve learning in subjects with neuro-development disorders, it may be essential to modulate the horizontal or vertical layout (within larger common areas/wards) of the *learning units* (classrooms) and related spaces described above, to accurately plan the color palette

and the other sensory stimuli such as noises (with the possibility of specific soundproof spaces), the lightness, with possibility of varying not only the brightness of the environments but also the ways of presenting the light.
As with typical developmental subjects, even for students with neurodevelopmental disorder, the organization of learning spaces will need to vary according to age: even in adolescence, the modularity of spaces should be guaranteed, ensuring both the sharing of experiences and the individuality of acquisitions and consolidation. Spatial and temporal variability of spaces, on the other hand, can often play an important role in preventing stigma (avoid identifying a "*room of the freaks*").

The Summer School in Alghero

The principles and modalities described above derive from the application to the clinic and to the teaching of neurobiological evidences derived from clinical research studies using neuropsychology, genetics, biochemistry, electrophysiology and neuro-imaging approaches (please refer to "*suggested readings*"). Participating in a summer school with a faculty characterized by different professional backgrounds and scopes (architects, but also psychologists, child psychiatrists and teachers) was particularly exiting since it was evident that although many concepts and constructs have become common to different professions, the intense and stimulating exchange of information on both theoretical principles and the practical experiences of the various disciplines is still crucial for the general welfare of pupils.

References

- Banschewski T., Zuddas A. (2015), et al, *Attention Deficit Hyperactivity Disorder*, 2nd Edition, Oxford Psychiatry Library, Oxford University Press, pp. 1-123, 2015
- De Rubeis S., He X. (2015), et al. Synaptic, transcriptional and chromatin genes disrupted in autism *Nature* 515: 209-2015
- Faraone S.V., Asherson P. (2015), et al. Attention-deficit/hyperactivity disorder, *Nat Rev Dis Primers*. 1:15020
- Lai M.C., Lombardo M.V., Baron-Cohen S. (2014), Autism, Lancet 383: 896-910
- Stichter J.P., Riley Tillman T.C., Jimerson S.R. (2016), Assessing, understanding, and supporting students with autism at school: Contemporary science, practice, and policy. *Sch Psychol Q*. 31:443-449
- Thapar A., Cooper M.. (2016), Attention-deficit/hyperactivity disorder. *Lancet*. 387:1240-50

4.2

Technology in the innovative school

Paolo Garau*

This work presents the process of study and analysis carried out in order to merge technological elements with teaching in an innovative school.
It is the project for the construction of the Up School, a private primary school, opened in Cagliari (Sardinia, Italy) in July 2015, stemmed from its principal's experience in the management of preschools. Its aim is to extend teaching and education to the concepts of wellness and balance. The entire project is based on the premise that underpin the creative model of a new school, introducing specific technological instruments to enhance the impact of the educational process. It is important to point out that the technological planning phase was initiated in parallel with the organization of the spaces and the teaching activity, in order to create a continuous system across disciplines to achieve the best possible result.

Innovation generated by three key elements

As I have mentioned above, the innovation in the Up School is based on three key elements, which are:
- Space;
- Educational model;
- Technological instruments.

The combination of these three elements created the real synergy between design and operativeness, which made clear that the school was (and is) truly innovative on many aspects, and also an example of detachment from traditional educational models. The Up School is a place where space becomes the guide to the spontaneous behavior of being together, allowing the children to use the environment in the most natural way without any kind of constraint that may lead them to a loss of concentration. The educational model itself is inspired by the northern European countries, eliminating elements of separation between teachers and pupils, such as the teacher's desk, and promoting discussion and game activities. This being said, a significant technological footprint was given to the whole school, and the features of innovation and diversification make it stand out among other local schools.

Which technological instruments have been chosen

The choice among technological instruments focused on equipment capable of attracting the children's attention by playing This is the key to educational success especially when the teaching activity is particularly demanding and difficult. During childhood, attention is captured by games, therefore a set of instruments capable of generating feelings of wonder, fun and curiosity has been chosen. It

* Paolo Garau, ICT expert and new technologies for learning.

has been observed that recreational and fun activities can reduce the barriers in learning, and that playing and new experiences make it easier to teach children the educational syllabus.

Based on these very simple and practical concepts, the following technological instruments have been chosen:

Interactive whiteboards and short-throw projector. They are very interesting tools for the teacher-led classes because they allow the children to develop manual agility and movement control through digital pens. These particular devices are directly connected with the computer, creating a large multi-touch screen, so that the users may have total control of the desktop.

Interactive tables for children. They are suitable tools when it comes to teaching science and geography, and they fit to younger children's needs, as easily accessible. They consist in a computer with a multi-touch screen, therefore they can include many children at once during the activities.

Passive 3D Projectors. Although the simple projection activity is very interesting in classic learning patterns, we wanted to create a space of relaxation without chairs where children could attend the 3D projection (with 3D glasses). This room is particularly appreciated during science lessons.

Interactive carpets with floor projection. This equipment represents a real gaming device for the youngest. Due to its multiple motion detection system, it allows children to perform group games on a wide work space.

Criteria in choosing technological instruments

During the planning phase, a lot of time had been devoted to the definition of the most suitable technologies to be integrated in the educational activity and therefore to its completion. It should be noted that the choice of the right technology had been established by two main aspects, namely, cost and usefulness for educational purpose.

The first criterion, although merely material and easy to overcome, shows that investments in the teaching area should be always weighted trying to achieve the best possible results with a cost that keeps the price-quality ratio constant, with the former as low as possible.

The second criterion, which is the educational benefit, is clearly dictated by the teachers' educational objectives and the educational model of the entire school. In order to put the criteria together, the teaching equipment has been chosen on the basis of the programmability and reuse of the devices in multiple contexts. This solution is the only one that can slow down the process of obsolescence of the equipment and maintain a continuous development of teaching, and, last but not least, it allows us to reuse the electronic device in different educational syllabi.

The technological footprint in the educational syllabus

Even though all the equipment mentioned in the previous points has been discretely integrated, there are some aspects that deliberately stand out to better characterize the educational offer of the school. For this reason, the school has opted for the integration of a baby FabLab, in which 3D printers are available, so that all the children can become little makers. The design activity is focused on a group game pattern, allowing children to experiment the following steps:

- the design phase, that follows a group discussion about what they must make;
- actual creation of the selected object;
- analysis of the result.

These steps are actually at the basis of the design of any experimental activity, which will be perceived as a game by every participant.
Furthermore, in addition to the baby FabLab, a coding lab has been installed, where children can acquire problem-solving skills by breaking each problem up into smaller issues. These are the first examples of creating useful algorithms to solve a problem.

Environmental control technologies

From the earliest designing stages, it was necessary to address the technological instruments to teaching as well as the environment and the children's safety.
As for the environmental control, the introduction of basic home automation (domotics) elements has been necessary, as they allow to tailor the learning spaces making them more welcoming and comfortable through climate and lighting control system.
Regarding environmental safety, a child localization system based on IBeacon tag was used; it makes sure that teachers know where each child is at any time. This kind of equipment generates alarms should any critical condition occur, and also gives constant information regarding the children's habits, allowing to shape the educational activity based on these features.

Example of a classroom in UpSchool , Cagliari.

BabyFabLab

4.3

A school that changes to meet the needs of each and everyone

G. Filippo Dettori*

In the last ten years, the Italian school system has significantly changed in terms of education, instruction and organisation. Up to a fairly recent past, the different educational institutions were organised to respond uniformly to the literacy of younger generations. Thanks to this, these younger generations were able to come into contact with knowledge and to acquire a cultural background. In the late 1970's, the school system began a gradual and evolving process that brought schools from being a strict educational and institutional framework to an educational and learning environment. Here students were able to express themselves and learn, all the while taking into consideration their own resources and their own limits.

The first revolution: Law 517/1977 and Law 104/92

The first important change of this evolving process was Law 517/1977. It recognized an important right: All children and young boys and girls must attend the same scholastic institution; those who formerly were "restricted" to "specially-made" institutions like differentiated classes or special needs schools, were finally admitted into mainstream schools and were now supported by specialized teachers. The important right of every child to experience a school life in the same educational context was therefore recognized. It offered each and every student the possibility of growing while being in contact with other students' differences. The implementation of this law provided the following: Classes with lesser students (article 7), a special needs teacher (article 2), the support of professionals from local institutions who would provide help to teachers with the preparation of a customized action plan (article 2). After the enactment of this law, and after the initial concerns of many teachers who were not ready to welcome students with disabilities into "their" classes, a long period of deliberation and didactic trial began. These deliberations and trials were determined to meet an objective that called for a rethinking of the school and to provide answers to questions regarding the educational needs, even of students with greater learning difficulties. With Law 104/1992, the Italian State tried to protect students with disabilities and provided a series of interventions with the following objectives: Early and quick diagnosis, adequate rehabilitation and care, academic, work and social insertion and integration. Clause 5 of the law, calls for a *Personalized Education Plan* for students with disabilities and calling for the collaboration of teachers, parents and health workers. It recognizes the importance of the school taking charge of students with disabilities and who should be supported in several aspects: Education-wise, didactically, medically and health-wise.

*. G. Filippo Dettori, Associate Professor of Special Education, University of Sassari.

An important contribution from the Word Health Organization: The ICF

A document which has a significant cultural impact in regards to recognizing the rights of people with disabilities, is the International Classification of Functioning, Disabilities and Health (IFC) published in 2001 by the OMS. Classification went from the term "handicapped" as declared by the OMS (ICDH 1980) to "a person with a disability" and therefore surmounting a model of intervention that was mainly medical and opting for a model that was both medical and social. The ICF changed its former position that required people with disabilities to adapt to an environment designed for so-called able-bodied people, thus acknowledging the need for social change in terms of perception of health, illness, disabilities, and unease. In terms more strictly related to education and learning, the ICF introduced great cultural changes and reminded schools and all learning institutions that a health condition could be more or less disabling depending on the environmental solutions. According to this vision, a disability is no longer a way of being of the individual but rather an encounter between an individual and a situation. It is a drawback that could be more or less important according to the relationship that is established between the person and the environment in which he/she interacts. Schools must primarily be sensible to health differences which, if not recognized, could result in exclusion and marginalization. It must, therefore, educate and sensibilize all citizens to commit so that a handicap does not become a source of discrimination and inequality. In educational contexts, it is important not only to promote the acquisition of knowledge but also to promote the global development of the person, offering everyone an instrument and enabling them to be autonomous in his choices and aware of the decisions he/she is making. In this regard, Morin reminds us that : "The objective of education is not to give the student an even larger amount of knowledge but to establish in him a deep inner state, a sort of popularity of the soul that guides it, in a tangible way, not only in childhood but throughout life " (Morin, 2000, page 45). At school, therefore, we must *learn to be*, or rather to become more and more aware of one's strengths and of one's limits. For this reason, a classroom as it was percieved before 1977, was not inclusive since it marginilised the so-called "unhealthy". However, over the years, classes have become more inclusive to offer a more flexible context and able to adapt its educational activity to the needs of each student.

The school of Special Educational Needs

The effort towards inclusion has become more intense in the last ten years and particularly with the recent recognition of an effective training that takes into account those with Special Educational Needs (SEN). The "Special Educational Needs" (SEN, BES in Italy) was recognized in Italy with the enactment of the Ministerial Directive of December 27, 2012 *Instruments of intervention for pupils with Special Educational Needs and territorial organization for school inclusion.* This directive asks teachers to implement radical changes in didactic planning that must respond to the plans of the Personal Education Plan (PEP) and to the needs of students who are bearers of problems or disorders that hinder their learning. The document states: "There are students requiring special attention, for a variety of reasons, in every class. The need for special attention may be due to social and cultural distress, specific learning and or developmental disorders, difficulties arising from the lack of knowledge of Italian culture and language because they belong to a different culture". Therefore, the request is to personalize teaching in order to "reach out" to the more vulnerable students and who have, in the past, often not found adequate solutions and have often precociously dropped out of school believing it to be less suited to them. This is exactly what occurred to Specific Learning Disorder (SLD) pupils, who after numerous failures, great frustration and sometimes humiliation, gave up on studying (Dettori, 2015). Even Law 170/2010 recognizes the right of a dyslexic, dysgraphic, dysortographic, and dyscalculic student to have educational support from teachers who must facilitate his/her learning. Furthermore, the law requires that teacher support avail of compensatory tools and dispensatory measures. With the Directive on SEN of 2012, it is assumed that by simply recognizing the difficulty that the student encounters, be it temporary (For example, a foreign student who has just arrived in Italy and does not speak Italian well), or permanent (For example, a student with SpLD), it is possible to promote their school inclusion. However, beyond the good pedagogic and legislative principles students with SpLd have not always received the necessary support from schools which leads to academic success. This was demonstrated by numerous studies (Pavone, 2015). Nevertheless, we must acknowledge that in the last few years students with SpLD have benefitted from a learning environment that is more inclusive compared to that of the past. This mainly thanks to the greater professionalism of teachers who are able to personalize teaching by paying closer attention to individual differences.

References

- Dettori F. (2015), *Ne asino né pigro: sono dislessico, Esperienze scolastiche e universitarie di persone con DSA*, FrancoAngeli, Milano.
- Dettori F. (2017), *Una scuola per tutti con la didattica per competenze*, FrancoAngeli, Milano.
- Morin E. (2000), *Una testa ben fatta*, Raffaello Cortina, Milano.
- Pavone M. (2015), *Scuola e Bisogni educativi speciali*, Mondadori, Milano.

4.4

Disability, learning difficulties and environment: Psychology of disability and Neuropsychology as basis for the development of a "School for all"

Donatella Rita Petretto*,
Paola Piras**, Ilenia Pistis**,
Carmelo Masala**

Since the 70s, in Italy there has been an huge cultural drive aiming to reduce special schools and special classes and to promote inclusion of students with various difficulties (intellectual disabilities, motor difficulties, sensorial difficulties and others) in the mainstream classes and in the mainstream schools. In the first period, the interest was addressed only to students with mild disabilities (law 188/1971), then with the law 104/1992, all students with disabilities, with any level of severity, were guaranteed the right to study and the right to study in mainstream schools.

Then there was not only an interest to assist people with disabilities but also to promote inclusion, to promote the rights of people with disabilities, and to guarantee them the higher level of autonomy and to promote inclusion in mainstream classes for each student, in any level of schooling and also in the university. Moreover, most of all, the right of schooling in the mainstream schools was ratified by law. With law 170, for the first time there was also a specific interest to learning disabilities, not mentioned in previous laws.

In the 70s, this sensibility towards these issues became widespread, and mainly as a sensibility towards "excluded people", and was devoted also to different aspects of daily life, environmental ones and family ones, and the consequences were the development of a lots of strategies for inclusion. Regarding environmental factors, strategies aimed to create inclusive environments: in the first step, there was the tendency to modify mainstream environment with "special" modifications, and in a second step, there was a tendency to create environments that were inclusive since birth. This field of study, Ronald Mace, an architect and product designer, and the director of the Center for Accessible Housing of North Carolina State University, developed a new approach named "Universal Design". In 1985, Ronald Mace proposed Universal Design as "the design of product and environment usable by all persons, in the wider possibility, without the need of adjustment and special assistive technologies"; a new vision on design addressed to the creation of environments and products "accessible for all" (Clarkson & Coleman, 2015).

* Department of Pedagogy, Psychology and Philosophy, University of Cagliari.
** Neuropsychological Medical Office Masala, Petretto, Piras, Pistis, Selargius, Italy.

In 2002 David Rose and Anne Meyer of the Center have developed a similar approach for Applied Special Technologies also for learning environments, named "Universal Design for learning": instead of creating "special" learning environments in mainstream environments, the aim is to design and produce inclusive learning environment since their birth (Bernacchio et al., 2007, Smith & Lorey, 2017).
According to the previous discussion, in this paper we aim to discuss briefly some hint of analysis, discussions, and some different conceptual models, and we discuss about the possibility to use them as basis for the design of inclusive learning environments. We describe the contribution of conceptual models of disability, the description of the heterogeneity of the students in the classes, starting from the clinical analysis of Special Educational Needs, the analysis of various cognitive functions, which could be impaired in people with learning difficulties. Last, we will discuss the role neuropsychology could have on discussion about the relationship between environment and learning difficulties (in positive for the enablement of students and in negative for the disablement of students).

Conceptual models of disability

In a first phase, discussion on and intervention in the field of the right to study of the students with disability were based on the first model on this issue developed by the World Health Organization (named ICIDH International Classification of Impairment, Disabilities and Health). In this model, authors described disabilities as difficulties in doing activities of daily living (ADL, with referent to both basic ADL and instrumental ADL) (WHO, 1980). In the evolution of conceptual models of disability, there has been a convergence of different international models on the idea that disability could be the consequence of a dynamic process. The reference to a dynamic process (and not more an attribute of the person, related to the relationship between the person and the environment) was stated clearly in the second model developed by the World Health Organization, named ICF (International Classification of Functioning, Disability and Health)) (WHO, 2001, Masala and Petretto, 2006, 2008a and b)[1]. This change represents an important modification of paradigm: from the making of the person responsible, to the study of the functioning of the person, the consequence of the interaction between the person and the environment. The change is from a point of view according to which disability is an attribute of the person to a point of view according to which disability is the consequence of the "negative match and impact" between the person and the environment[2]. With a focus to the school context and to learning as a whole, the focus changes from the students with his/her characteristics to the interaction between the student (with his/her characteristics) and the environment (with its characteristics, seen in the wider meaning as physical and social ones). In that sense, it becomes important a deeper knowledge of the characteristics of the student but also the characteristics of the learning environment. However, it is even more important a deeper knowledge of the real and possible interaction between these two elements (Masala e Petretto, 2006, 2008a e b, Petretto, 2016, Petretto, 2017).
Also the change from "insertion", "integration" and "inclusion" in this field of study has not only a terminological value, but the semantic choice refers to a model of approach and an anthropological vision which underlines the importance of the interaction between the person and the environment and between the different people (people which learn and people which teach, also).

---

[1]. In the Italian context, law 104 of 1992, which is an important reference for the right to study of people with disability, proposed a definition of "handicapped person" with a reference to ICIDH model. Only later, the dynamic vision described by ICF was introduced.
[2]. See also Masala and Petretto (2008 a and b).

From disability to Special educational needs: a neuropsychological analysis

The new interest in the Italian context to different kinds of "special educational needs" extends the number and the heterogeneity of clinical conditions and other conditions related to learning difficulties. The main consequences are the awareness of the heterogeneity of the students in the classes. Even in the past, it was enough difficulty to list all the clinical conditions described by the law 104 of 1992, but this is only one of the subgroups described as special educational needs, other subgroups refer to Specific Developmental disorders, borderline intellectual functioning and other socio-cultural and linguistic based conditions.

The actual challenge in the school and in the clinical field is to understand this heterogeneity (Petretto et al., 2017). What methodological instruments could we used to understand it? What kinds of disciplines could help? If the Psychology of Disability gives us a general theoretic framework aiming to describe the dynamic processes of interaction in the study of human functioning, neuropsychology and school psychology could give instruments to study in a deeper way individual differences and, again, dynamic processes of interaction[3]. A Study of human functioning based on the study of the cognitive functions involved in learning[4]. International literature in this field is very rich of studies on cognitive and neuropsychological functions related to learning. There a lots of researches on working memory, long term memory (implicit memory and explicit memory), attention (selective and sustained), linguistic abilities, praxic, visuospatial abilities, problem solving abilities (Giofrè et al., 2016, Smith-Park et al., 2016, Petretto et al., 2016, Liebel e Nelson, 2017, Petretto e Masala, in press, Petretto et al., in press). Executive functions and their relationship with learning is nowadays a keypoint in this field, mainly during adolescence and adult life, where they seem to have a crucial role.

There is also another field of research interests, related to learning abilities and the difference between basic and instrumental learning abilities, in different phases of the learning life. The study of their relationship with different environments and how learning abilities interact with different learning environments is the next step toward the creation of inclusive learning environments and, possibly, toward the "Universal Design for learning" (that we have previously mentioned).

For sure, it is not a simple field of study, it requires the description of a complex interactive model in a complex framework and it implies a complex circular cause relation. The model is based on the universality of ethical values and equal opportunities and it promote the acknowledgement of the individual differences and the real accessibility of the learning processes. Nowadays, we are still far from a widespread use of this knowledge in the school, even if scholars are creating a network of experimental knowledge and intervention in school, and our daily work aims to this goal (Bernacchio et al., 2007, Smith, Lowrey, 2017, Petretto e Masala, in press).

---

[3]. Perhaps a contemporary new way to study cognitive functions related to learning is based on School Psychology which assesses cognitive and neuropsychological functions by standardized tests and uses theoretical models like the so-called CHC model of Cattell, Horn, Carroll and the "Cross-battery assessment" methodology (Mc Grew, 2005, Mc Grew et al., 2010).

[4]. It does it not mean that we aim to exclude other functions, like affective and social ones. We are aware that there is close relationship between cognitive functions and other functions, mainly in some clinical conditions (see for example some descriptions of "nonverbal learning disorder") (Mammarella e Cornoldi, 2014).

References

- Bernacchio C., Mullen M. (2007), *Universal Design for learning, Psychiatry and Rehabilitation*, 31(2), 167-169.
- Clarkson J.P., Coleman R. (2015), *History of Inclusive Design in the UK. Applied Ergonomics*, 46, 235-247.
- Giofrè D., Stoppa E., Ferioli P., Pezzuti L., Cornoldi C. (2016), *Forward and backward digit span difficulties in children with specific learning disorder*, Journal of Clinical Experimental Neuropsychology, 38(4), 478-486.
- Liebel S.W., Nelson J.M. (2017), *Auditory and Visual Working memory functioning in college students with attention-deficit/hyperactivity disorder and/or Learning Disabilities*, Archives of Clinical Neuropsychology, 7, 1-12
- Mammarella I.C., Cornoldi C. (2014), *An analysis of the criteria used to diagnose children with Nonverbal learning disability* (NLD), 20(3),255-280.
- Masala C. Petretto D.R. (2006), *Disablement, assistive technologies and computer accessibility: hints of analysis through a clinical approach based on the ICF model*, in Minati G., Pessa E., Abram E. (eds): Systemics of Emergence: Research and Development, Springer, 169-177.
- Masala C., Petretto D.R. (2008a), *From disablement to enablement: conceptual models of disability in the 20th century*. Disability and Rehabilitation, 30(17):1233-44. Review.
- Masala C., Petretto D.R. (2008b), *Psicologia dell'handicap e della riabilitazione: dal "hand in cap" alla partecipazione*, Ed. Kappa, Roma.
- Mc Grew K.S., Wendling B.J. (2010), *Cattell-Horn- Carroll cognitive-achievement relations: what we have learned from the past 20 years of research*, Psychology in the School, 47(7), 651-675.
- Mc Grew K.S. (2005), *The Cattell-Horn- Carroll theory of cognitive abilities*, in Flanagan D.P., Harrison P.L., *Contemporary intellectual assessment: Theories*, tests and issues, 2nd edition, New York, Guilford Press.
- Norwich B. (2016), *Conceptualizing Special Educational Needs using a bio psychosocial models* in *England: the prospects and challenges of using the International Classification of Functioning framework*, Frontiers in Education, 1:5.
- Petretto D.R., Bariffi F., Jimenez E., Volterra S., Masala C. (in press), *I Bisogni Educativi Speciali: il diritto all'istruzione in una prospettiva inclusiva*, Edizioni Jovene.
- Petretto D.R., Masala C. (in press.), *L'intelligenza: tra psicologia clinica ed intervento*, Ed. Aracne.
- Petretto D.R., Piras P., Pistis I., Tradori E., Camboni M.V., Staico F., Palmas F., Lussu C., Preti A., Masala C. (2016), *Difficulties in reading, learning difficulties and neuropsychological profile* on WISC-IV in Italian children, IEEE International symposium on Medical Measurement and Applications, 978-1-4673-9172-6/16.
- Petretto D.R. (2017), *Inclusion and university: a key point in the assessment of the quality of the services provided to the students*, Ministry of Education of the Republic of Belarus, Minsk.
- Petretto D.R. (2016), *Inclusion in school and university: services provided* in the State University of Cagliari, International Conference "Inclusive processes in education", Ministry of Education of the Republic of Belarus and Belarusian State Pedagogical University named after Maxim Tank, Minsk – Republic of Belarus, 27-28 October 2016 (invited talk).
- Petretto D.R. (2016), *Percorsi inclusivi e bisogni educativi speciali: Scuola ed università*, Convegno "Sui Generis 2016. Le giornate Cagliaritane delle Pari Opportunità", Ordine degli Psicologi della Sardegna, Ordine dei Medici della Provincia di Cagliari, Ordine degli Avvocati di Cagliari, Cagliari 8-9-10 settembre 2016, Polo Economico Giuridico dell'Università di Cagliari (invited talk).
- Smith S.J., Lowrey K.A. (2017), *Applying the Universal Design for learning framework for individuals with intellectual disability: the future must be now*, Intellectual developmental disability, 55(1), 48-51.
- Smith-Spark J.H., Messer H.L.A., Messer D.J., EdvardsDottir E., Ziecik A.P. (2016), *Executive Functions in adults with developmental dyslexia*, Research in Developmental Disabilities, 53-53, 323-341.
- WHO, World Health Organization. *International Classification of Functioning: Disability and Health*. Geneva: World Health Organization, 2001.
- WHO, World Health Organization. International Classification of Impairments: Disabilities and Handicaps: A manual of classification relating to the consequences of diseases, published in accordance with resolution WHA 29.35 of the 29th World Health Assembly, May 1976. Geneva: World Health Organization, 1980.

Experiences of research and project:

The school and its surroundings. Generosity and sincerity

Simonetta Cenci, Alfonso Femia*

The experience of the Summer School of Alghero allowed to demonstrate some things that in reality are little known, or at least not perceived, and consequently do not make critical mass together with other cultural, theoretical, operational, social, and project experiences making the school world more in crisis than what actually is, at least in its content.

In the Summer School, for those who did not know or had never paused to reflect, it was shown that the school is many things and each one is important, necessary, fundamental, foundational. In the various dialogues and projects comparisons, a new shared sensibility has emerged: the fact that the school is not trivially or simply a building, good or bad, is not an aseptic and independent volume, it is not a service function and nothing else.

The school is a vast world, which must be faced by different directions because the project is mainly based on a vision, both real and imaginary, and then in shape and language, responsible and dreamlike, where these apparent contradictions are instead approaches that they complement and feed each other in order to make the school a place to learn, discover and study the world and life without closing doors or without a curious gaze.

We learned a lot during the ILS Summer School thanks to the different contributions, to the positively "factious" dialogues, to the different points of angulation from which the school is faced, personally discovering a series of people "rich" in content and generosity, where, with a broad and open vision can be glimpsed an idea of planning in each of these actions.

The project work has been stratified through the look to the city and its territory, listening to the different experiences that have taken place, the deep involvement of the participants.

We have discussed and found in various projects how the school can and should be the device for the relationship of urban spaces that anticipate the school itself, surround it or perimeter it, how they build it from the inside. The school for us, for our work group, has defined itself in a plot, a sequence, a rhythm between public spaces, spaces of relationship, spaces of perception, choral and individual dimensions, "intimate" and collective. It was understood and made to understand how a small nursery school can reconfigure places initially conceived without urban design, without the will or ability to make them "public" in everyday use, in the collective dimension of a city. With the project group we have proposed to proceed with planning according to two acts that we

* Alfonso Femia and Simonetta Cenci, Atelier(s) Alfonso Femia.

find in the didactic experience of the school: reading and writing. The reading of the area unfortunately shows what is often found around schools or school complexes built in recent decades, an area without real urban connections neither collective relations (public spaces) nor "intimate" (protected gardens). The school even before being a building is "the way to get to school", stands in front of the school, expects the entrance and stops at the exit, where can be experienced moments during the day between inside and outside, where is possible to discover, perceive, use the neighbouring "landscape" as a formative, cultural and recreational resource. It there that we must pay the most attention. It is in the creation of that "place" that the school roots in young people, in us, the civic, public, of belonging, of identity, of the will to grow, confront, study ideas.
It was therefore clear and shared with all the architects of the project team how the theme of the path and the void was the themes to be addressed to give an answer that resolves the existing fragmented drawing, or better, the fragmented non-drawing.
The path becomes a road, becomes a square, becomes a perceptual device, becomes a fence, a porous system and a permeable opportunity. Enter into dialectic with the rediscovered and recreated plot of the trees that host playground areas connected to it among olive trees and fruit trees. The void becomes a project opportunity through the possibility to trace a line, the path, and to create a "surrounding" action through it. The goal is to create a sense of belonging to the place, to foster relationships, the meetings seeking a spatial "unicum". Wire, line, path. Fence, area, surrounding. Geometry, plot, nature. The nursery school thus becomes the instrument, the joint of relationship, which builds a pedestrian porosity, anticipates the existing school, creates a relationship of landscape. It is a founding act of multiple responsibility: as a place of formation, as a place of civic and social culture, as a place of relationship, as a place of "design" and belonging.
Beside the project and reflection experience, that it brought and shared with the students and the architects, there was a strange magical convergence of concepts, words, points of perspective between the different actors of the workshop that can not be not remembered as a precious and "particular" element of this summer school conceived and created by Massimo Faiferri.
The search for weak territories, the search for "leftovers" to connect them, is a different modality of Joao Nunes to reach the same goal previously told about the need for a work on the place made of anchoring and belonging, responsibility and vision.
We lovely have been talking for over ten years about architecture as a necessary act of generosity, which must be declined under different profiles, moments, and with the various interlocutors, as well as being a project.
In Alghero we saw on stage another fundamental and complementary value to generosity, for us who have a "south" soul: the sincerity.
Many things have been said and done, but this is a time without memory, looks, without courage, vision, pleasure. A decadent and mediocre time, even if extraordinary, as a passage between different times. Let's be both responsible and bold, dreamers and pragmatic, visionary practitioners of reality.

NEW SCHOOL COMPLEX

Authors: 5 + 1AA Alfonso Femia, Gianluca Peluffo Simonetta Cenci, with Diego peruzzo, Sandro Cavalieri
Client: Municipality of Zugliano
Architects: 5 + 1AA with Diego Peruzzo, Sandro Cavalieri
Client: Municipality of Zugliano
Year: 2010-2013
Surface: 2500m²
Photography: © E.Caviola

Designing, building a school is a public event. Therefore, we imagined a building that was both repesentative and functional, recognizable and simple, endowed with an ability to dialogue with the territory and the landscape, but protected and protective. The project stems from two territorial elements: the horizontal line of the area; the profile of the mountains and hills. The theme of the project is to exploit these conditions in order to build a significant and welcoming building, that is, a system capable of declaring its belonging to the context, strengthening its characteristics

The patio space

The great coverage and the landscape

The patio as a relationship space

The patio as a space for educational activities

The school entrance

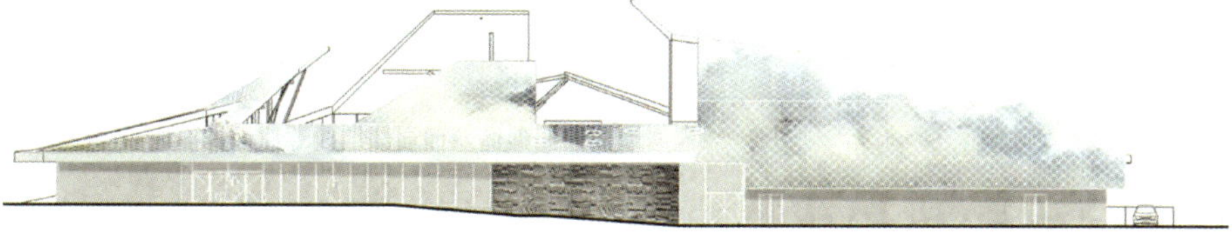

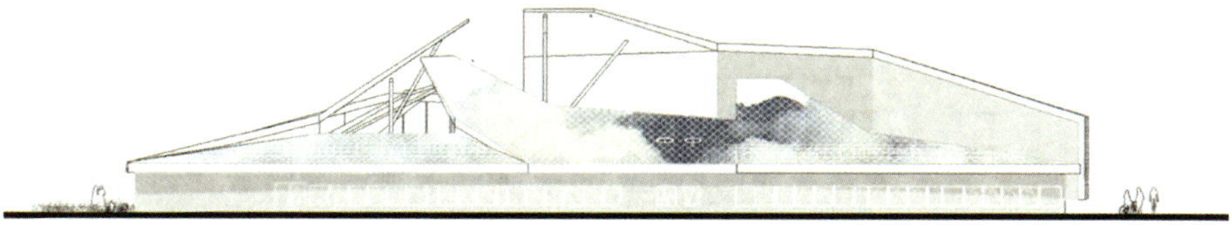

South and west elevations

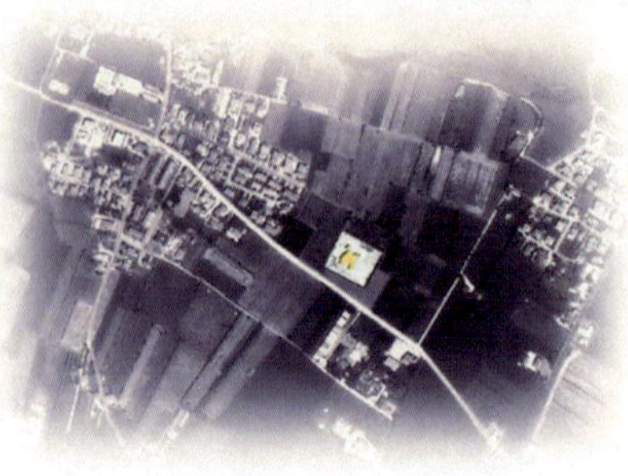

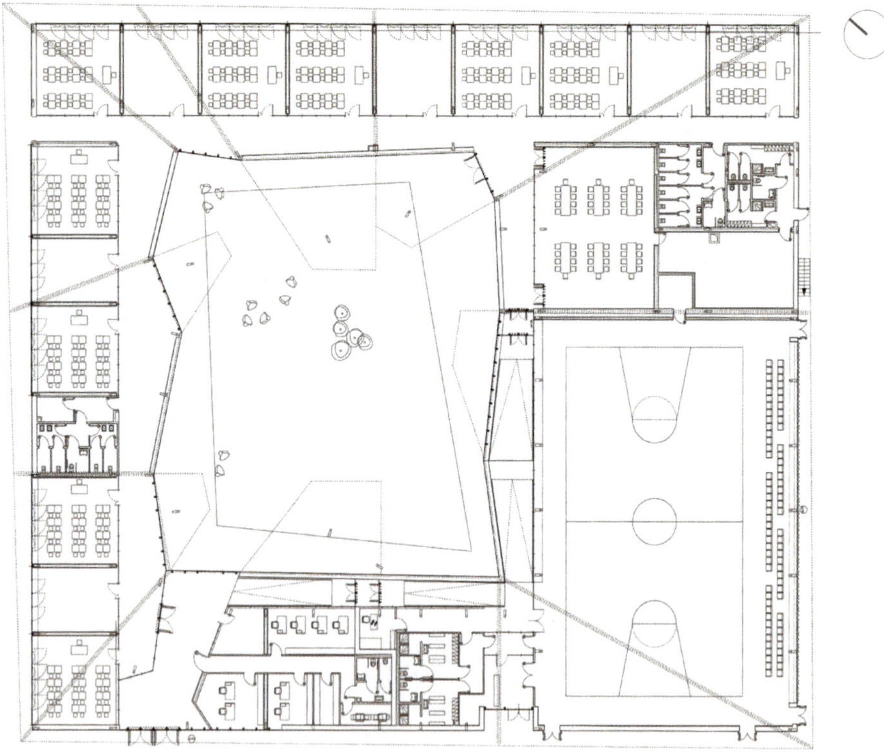

Distribution of learning spaces

The dialogue between architecture and pedagogy: MoDusArchitects

Samanta Bartocci*

MoDusArchitects' work and research experiments show deep reflection on the harmony, sometimes fleeting, of creating architecture. When we speak of architecture, it should be spontaneous to think that environments are constructed for those who will live in them, and when we speak of space and learning environment, we are moving through the uniting relationship of dialogues between architecture, pedagogy and the narration of our times. As with poetry, the space of architecture actually helps man to practice with symbols and continually test them. Sometimes they are encrypted codes and not conventional signs, so to interpret and decipher them is already a production of space. Therefore every action on space has an inevitably social dimension as, whether we like it or not, a different distribution of inhabitants corresponds to each choice and the expressive matrices of the volumes in architecture, in their turn, shaped and sculpted, adapt to what the project finds.

Basing ourselves on the logic of an active space, therefore, that will accompany learning environments as spaces of complex interactions will deeply affect the quality of learning outcomes. In this sense the school may find even more urban relations, as a space of everyday and not as a "machine to educate", and may found the pre-requisites for setting up spatial alternatives to the traditional school model, by seeking other possibilities originating from an attempt at an interdisciplinary dialogue between architecture and pedagogy and from an attempt at promoting the creation of stimulating learning contexts. *"To design a school, and not simply build it, is a gamble that can never be taken for granted. It is an open-ended game. An unending succession of acts. It is a project for time to come. In this sense it remains infinitely open and unfinished, and is based fundamentally on faith in the extraordinary potential of awareness and responsibility as an expression of the commitment of managerial teachers, architects and administrators towards the creation of the new places of culture for the community." (Weyland, Attia, 2015)*

These themes are at the centre of attention of the MoDusArchitects Studio, which thanks to an international planning competition has experimented with and created for the Municipality of Bolzano, the architecture of two new school buildings and a square in the new Firmian district, on the edge of the city. The Infant Complex, composed of the crèche, infant school and children's centre was the first of the two buildings to be completed in September 2012, while the junior school and local library have recently been finished.

We have seen that planning educational space nowadays has to contend with a whole series of themes that permeate teaching and learning in the contemporary society. A variety of meanings and research converge in the themes: the entry into schools of new communication technologies, new cooperative learning methods, interdisciplinarity, the quest for new expressiveness and languages and

* Samanta Bartocci, DADU, University of Sassari.

the reciprocity of learning and teaching. With this MoDusArchitects Studio project it is already possible to see the effects of this change in learning methods on urban experience. The spaces of the city are meant as relational places, educational spaces in which the representation of *opposing expressive forces* can be set up, each of which has the ability to free a great amount of energy opposing similar resistance, and aiming at equilibrium. In the two buildings the "opposing" forces and forms face each other in a dialectic *"[...] between generating order and variety of the parts"* (Scagnol, 2013), and their location pre-empts a central public space, an agora, a meeting space and the civic dimension of the district that extends towards the church and the parish centre. This dialectic between the solid geometric, right-angled shapes of the junior school and the curved ones of the nursery complex, tends toward an overall redevelopment of the district introducing the natural quality of the public park and surrounding countryside. The design of the square also interprets some artificial ground as a regular stone plateau from which green islands and trees emerge completely irregularly, dotting the great extension of public urban space.

The Junior School and Civic Library of the District

By the way it is arranged the junior school has set up a series of relations on an urban scale with the streets and future church of "Mother Teresa of Calcutta". The main entrances to the school and library open onto the side of the square, delimiting the urban space. The building was created in two sectors forming a double "U" shape: on one side common functions like the canteen and hall are set out towards the countryside, with the classrooms upstairs, while on the other, towards the square there is a covered educational space, but also an environment leading to the city, which includes a gym area, a space for information and communication and the local library.

The latter has direct, independent access from the square, but it can also be entered directly from inside the junior school, both on the ground floor and the first floor. A large glass window over the square amplifies the need for continuity between inside and outside, whereas a large curving building juts out towards the covering loft, providing natural light for the reading spaces. The building creates a spatial condition that offers the city an inclusive dimension, which as such includes the outskirts in the heart of the city, equipping them in fact with urbanity and a degree of participation in full urban life.

The crèche, infant school and children's centre

On a parallel and antagonistically, the Infant Complex contains and integrates three functions in a single building, crèche, infant school and children's centre, and was the first of the two buildings to be completed. In this account, the dialogue between the architects and various stakeholders took on an important role in the design phase, in which they sought *"careful attention to the different pedagogical approaches offered by the three facilities and identification above all of the importance of educational continuity for the child, with the educational course over the years being experienced in a fluid, natural way"* (MoDus Architects, 2014). In this combination some prerequisites are merged which will make it easy to consider school space and learning environments an "entity able to be planned further", and the history of the cooperative, shared project has acquired 'becoming' meanings, thanks to the action of the subjects involved. In this building separation and interaction are matters that cause the trend of the shapes to develop a sequence of convex and concave parts creating strong relations with the public park, but also with the three courtyards set out inside. The shape of the courtyards subdivides and crosses the building, giving various opportunities to the roomy spaces of distribution and the classrooms at the edge. The offer of a variety of internal and external spatial configurations stimulates in the child the 'sense of possibility', of exploring space and understanding what a safe place is. Outside the uncovered rooms there are play places and places for meetings and relations between teachers and children, where the hard work and sense of choice and freedom become combinable and active.

Design for learning space, or more specifically school design, cannot but consider the subjects as active elements in the space. For the space, as an engine of development of processes and actions, takes on a revelatory role towards its surroundings and spontaneously suggests its future evolution. Schools, in turn, reject the idea of being knowledge containers to be tapped into just in a reflective way, and can therefore take on the role of active spaces to stimulate experience, for which abilities are conceived as *"a bricolage of emerging knowledge and properties that are strictly tied to the contexts in which they mature and develop as the temporary, never finite, result of a learning course"*. (Tagliagambe, 2016)

The spaces of the game. Photo Hannes Maraner.

References

- Modus Architects (2014), *Polo per l'infanzia Firmian*, Bolzano, Scuola Elementare di Ora, in n.3 "Obiettivo Concorsi", Trimestrale di informazione dell'ordine degli architetti PPC della Provincia di Trento, Trento.
- Weyland B., Attia S. (2015), *Progettare Scuole - tra pedagogia e architettura*, Guerini Scientifica, Milan.
- Tagliagambe S. (2016), *Idea di scuola*, Antonio Tombolini editore, Loreto Ancona.

SCHOOL COMPLEX, ELEMENTARY SCHOOL WITH PUBLIC BRANCH LIBRARY

Authors: MODUS architects
Location: "Firmian" district, Bolzano
Client: Municipality of Bolzano
Realization: 2015
Architects: Sandy Attia, Matteo Scagnol

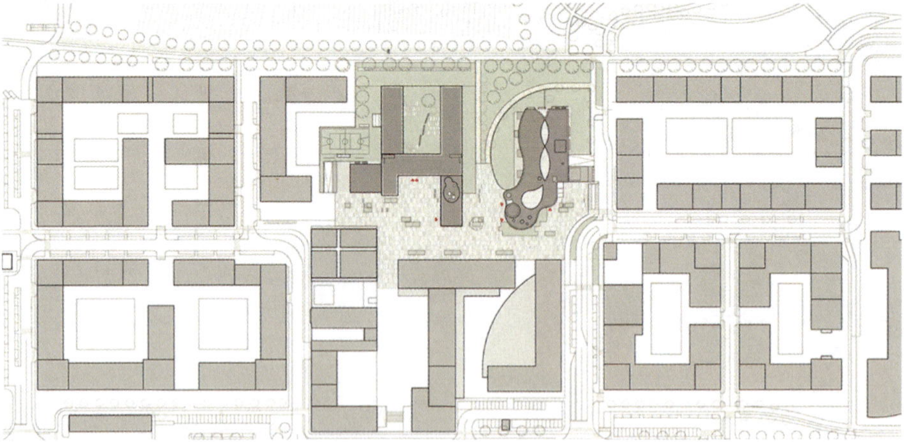

General plan

The neighborhood library, the main hall, the canteen and the gym, are spaces designed for the community, not just for the school center. Photo Oskar Da Riz.

The school and the library. Photo Oskar Da Riz.

PRE-SCHOOL, KINDERGARTEN AND FAMILY CENTER

Authors: MoDusArchitects
Place: "Firmian" district, Bolzano
Client: Municipality of Bolzano
Realization: 2012
Architects: Sandy Attia, Matteo Scagnol

The nursery school, kindergarten and child center, Bolzano.
The patio and garden for outdoor activities. Photo Hannes Maraner.

The learning environments

5 WORKSHOP

5.1 AH01
Gianluigi Mondaini, Lino Cabras
The School of "Sacro Cuore", Alghero

5.2 AH02
Alfonso Femia, Simonetta Cenci
Samanta Bartocci, Fabrizio Pusceddu
New micro-crèche for "La Pedrera"
Unified School N°2, Alghero

5.3 SS3
Massimo Ferrari, Claudia Tinazzi
Extension of the State Middle School N°11, Sassari

5.4 SS4
Sandy Attia, Matteo Scagnol, Francesca Arras
Elementary School N.11 and the residence for the elderly, Sassari

5.5 PT5
Patrick Fransen, Francesca Rango
Ex Elementary School De Amicis, Porto Torres

5.6 PT6
Valter Caldana, Sabrina Scalas
Playground and satellite district, Porto Torres

Introduction to case studies

The school as a common place

Samanta Bartocci*

This paper is a reflection on some of the terms in use in the vocabulary common to architecture and pedagogy, in an attempt to explore some meanings inherent in the educational system considered as opportunities, as experience of inhabiting and as a reflection of traces on the territory of an urban condition. The need to search between the lines and definitions is linked with an initial circumstance offered by the extension of the concept of inhabiting.

Much of the research on preadolescence still highlights the school as a common place, the representative nature of which is essential for education; at the same time it illustrates the critical modern condition. The school has begun to welcome the change, though in an uneven, rather disorganised, fragmented way, but is nevertheless the place where most effort is needed and where the change present is already responding with new forms, though not sufficiently reported. *Stories* and not the *standard* are addressed by the school, as Bernardini realised well with his "stories without an ending" - tales and fables without an end, to give young readers a way to invent a personal ending.[1] And with this need to be the authors of our own evolutive process we still find in the school, though especially in educational practice and the capacity to inhabit the school building, the idea and definition of protected, isolated place, separate from the contest, where education is, indeed, dispensed in safe circumstances, keeping out the others and the city.

In spite of this, the city, which has always in the past been the space of social and cultural integration, innovation and exchange, has shown methods in the urban policies of the last twenty years that were suspending the rights of individuals and their groups, retracing in society a sort of rhetorical trap, between the marketing rhetoric and the rhetoric of safety, which shape its evolution.[2] *"All the reference points that gave solidity to the world and favoured logics in the choice of life strategies (jobs, capacities, personal ties, models of convenience and decorum, concepts of health and disease, the values we thought should be cultivated and the approved ways of doing so), all these and many other reference points once stable seem to be undergoing great change. There is the feeling that several games are being played simultaneously, and that each one's rules change during the game.*

*** Samanta Bartocci, DADU, University of Sassari.**

[1]. From "Bacchette di Lula", by Albino Bernardini, ILISSO Ed.; Albino Bernadini was born on 18 October 1917 in Siniscola, Nuoro province, and wrote fables and tales for children, such as "Bobby va a scuola", "La banda del bolide", "Tante storie sarde", "Le avventure di Grodde", "Il palazzo delle ali", up to the very latest "Un viaggio lungo trent'anni". He has received more than 15,000 letters from children all over Italy.

[2]. A. Calaffati, La città degli altri, (Ibidem) N° 1 Le Letture di Planum – Planum – The Journal of Urbanism.

This epoch of ours excels in dismantling systems and melting down models, every kind of system and every kind of model, randomly and without warning." (Bauman, 2002) In spite of everything, therefore, if we consider the need of a dialogue between the outside and inside of school, and propose the principle of a reversal of perspective to support an opening and a way to oppose the separation scenario, then school spaces and learning environments - in such a dialogue – should open up towards some possible interpretations.

First of all, the school building could be defined as an open structure, a cultural, civic place of different dimensions. A space able to sustain relations with the territory and citizens, and able to forge a strong bond with the different processes of urban life. The city may, on the other hand, be an instrument of self-evident didactics (Brusa, 2012), however, and may increase the development of learning processes in the city and on the territory that are more and more intertwined and usable within an extended, spatially articulate system. Consequently, to consider the city as an *educational* environment entails being conscious developers of knowledge and interpreters of the progressive change in the way and relations constituting the new perspectives. To educate and at the same time educate oneself are positive dimensions of a single, 'becoming' process, constantly tied to cultural, social, political and economic changes.

If we reflect on the Latin origin of *to form*, the reference to the meaning "give shape to something" is clear. And through the forming process we know how to combine forming with forming oneself in a course of growth and constant change in which the subject-child-man is the active protagonist in taking shape throughout his life, and in his dynamic role in society.

In this sense, fuelling the natural capacities for action of each individual may represent an instrument able to foster the emergence of a process of attachment, accountability and awareness-raising on the part of the community towards places and the city. So terms like *environment, learning, new technologies,* but also *wellbeing, growth* and *belonging* are some of those key words that represent the need to change the geography and shape of the spatial configurations of design for learning space for *"inhabitants"*. A modern tendency is maturing, involving the belief in and importance of *architecture* for *learning environments* which however encourages the formation processes of sharing and cooperation for the wellbeing of the subjects involved. Pedagogical studies confirm this need, recognising that space is an existential, lived-in dimension, to the point of becoming an identity reference point in human experience, and indeed, studies on proxemics confirm the strong influence of space on people's behaviour when retracing, through empathy, the cognitive capacity of perceiving what is felt by another individual in a common context and entering into resonance with them, to the extent of determining their cognitive processes.

"This means that empathy and its consequences are not strictly biological, mechanical processes but are filtered through culture and its types and specificities, which determine different reactions to and perception of the same sensory stimuli. Therefore, cultural and emotional affinities favour and facilitate the empathy process and maximise its efficacy; on the contrary there are factors that hinder its stimulation, and among them are above all the individualist culture, competitiveness, evasion and hedonism. Empathy tends to be inhibited if the other is perceived as a rival, such as a competitor or if the reciprocal observation and interaction take place in a context of disinterest, lack of attention or of spectacularised exhibition, and devoid of actual emotional tension of one's feelings.

These results are a further confirmation of what has been said ... by Ivo Lizzola on the essential, irreplaceable function of the school as a "common place": the only institution in which the generations (and nowadays we may also say different forms of life and varied styles of thinking and cultures) encounter each other and communicate. This potential of the school, Lizzola has rightly observed, is used too little, whereas, if it were exploited with greater attention, it could constitute a powerful factor of development of that empathic fibre able to act as an effective remedy against the current, increasingly alarming flight from citizenship and the crisis of social ties." (Tagliagambe, 2013)[3]

In this *cognitive spiral* it seems important to consider that the space of action for the inhabitant in the "world" is essentially identified with that of design, understood as the transformation of an environment, a territory and, more generally, a place and condition of improved habitability. And if this were also true in the planning of the spaces where – temporarily – inhabitants live? Well, it would be a case of setting up relations between the internal dimensions and the different external ones, between the body of the building and the recognition of the existence of the space possible - and the design work.

An initial phenomenological recognition on inhabiting that runs through Heidegger and Merleau-Ponty, to arrive at Norberg-Schultz's theories, leads us to the possibility of interpreting architecture as *experiential appropriation* of the world. Hence the inhabitant obtains through the *genius loci* (spirit of the place) the possibility to inhabit and adopt space, especially everyday domestic space, as a mental as well as physical space. By using his imagination, man-child manages to find his own sense of belonging to the place. For Hannah Arendt, in inhabiting, to act is as fundamental as to build, in that both are ways of generating something new and modifying the undifferentiated space, in places.

In observing this composite background consisting of the relation subject-collectivity-environment and the construction of space starting from action, it is interesting to resort to the meaning of the *"term "environment", with its prefix "amb" indicating an organisation simultaneously single and double, in the sense that it indissolubly concerns two different, contrasting realities: an object, the physical and social context, and a subject, the individual, who moves within it and in the end becomes "part of its body". But we can also consider, when analysing the term environment, that the prefix amb derives from ancient Indo-European and gives all words beginning with amb a meaning that concerns 'surroundings', i.e. what lies around. Environment derives from the present participle of the Latin verb "ambire", which means to wander around, have relationships with the surrounding context; the environment is therefore, in this meaning of the word, what lies around someone or something, with which they have relations of reciprocity or exchange... So the environment is a space where life inhabits, it is where the fibre of mobile and transformational relations become established in an inextricable mix and insemination-contamination between nature and artifice, man and world."* (Berlinguer, 2008) From these and other reflections linked with the evolution of the concept of school building as a social object (Markus, 1993) and with the change in the *paradigm of teaching* towards that of *learning*, the link between environment and learning has been explored, reappraising for some time relations between education, community and territory, even though we very often notice a clear resistance to the change and cultural exchange between disciplines, contexts and interlocutors

---

3. To study this further, see Silvano Tagliagambe in "Competenze e ambienti di apprendimento: cambiare paradigma".

that notoriously do not communicate with each other but could begin to do so. (Weyland, 2014) In this perspective there is convergence between the dialogue with the ties suggested by the different spheres, and the benefit of creativity emerging from the experience of cultural diversity.

References

- Bauman Z. (2002), *La società individualizzata*, "L'istruzione nell'età postmoderna", Il mulino Bologna.
- Berlinguer S. (2008), Lo spazio dell'azione: Corpo e Progetto, PhD thesis, DADU, UNISS.
- Brusa A. (2012), "Dalla città come aula alla città-archivio. Note per un possibile curricolo verticale sulla città", available online at: www.comune.pv.it/museicivici/pdf/annali28/49%20Brusa%20Didattica.pdf
- Markus T.A. (1993), *Buildings and Power: Freedom and Control in the Origin of Modern Building Types*. Routledge, London.
- Norberg-Schulz C. (1979), *Genius loci. Paesaggio ambiente architettura*, Electa, Rome. Rizzolatti G.,
- Mascheroni, 29 October 2013. Tagliagambe S. (1991), *L'epistemologia contemporanea*, Rome: Editori Riuniti.
- Sinigaglia C. (2006), So quel che fai, Milan: Cortina.
- Tagliagambe S. (2008), *Lo spazio intermedio - Rete, individuo e comunità*, Milan: Università Bocconi Ed.
- Tagliagambe S. (2013), Competenze e ambienti di apprendimento: cambiare paradigma, Bergamo, Liceo
- Weyland B. (2014), *Fare scuola*, Edizioni Angelo Guerini Associati SpA, Milan.

Up School Sant' Elena, Up School Cagliari,
designed by Lino Cabras, Silvia Farris and Fabrizio Pusceddu,
photo by Stefano Ferrando

Centro Infantes Cagliari, project by Lino Cabras, Silvia Farris and Fabrizio Pusceddu, photo by Stefano Ferrando

Up School Cagliari, designed by Lino Cabras, Silvia Farris and Fabrizio Pusceddu

Distracted perception and the school as a misunderstanding

Fabrizio Pusceddu*

It is all a great misunderstanding. We look with admiration at north European schools, but close the doors of the classroom to avoid distractions. We want to design innovative schools, but basing ourselves on the regulations of 40 years ago. We think of open schools like urban centres, but put up fences for safety. We divide children's bathrooms according to sex, paint the walls with implausible colours in the name of theories anything but scientific, and cover the grass and ground with rubber. This is the actual situation into which the workshop's activities fit, proposing a series of themes on different projects, but all anchored to the same approach aimed at providing a contribution applicable to the debate on the forms and ways of designing learning spaces, starting with those of schools. Not salvific solutions, but methods and instruments that can be placed within a relationship of reciprocal complementarity between pedagogical needs and space design, both in the realisation of new school buildings and in the extension of existing ones or their upgrading or re-use, or even in imagining the school as an urban playground.

This means breaking up the traditional models of school architecture in defining contemporary (not necessarily innovative) learning environments that are dynamic and interconnected, and will contain features within them to stimulate knowledge, able to accompany a process aimed at constructing and enhancing abilities in a programmed, continuous manner (learning outcomes).

According to the CENSIS *"school drop-out is the expression of processes of resource dissipation (human and material) that are increasingly besetting the Italian education system, which does not seem, as a whole, able to respond to the educational needs of the young people of today."* The workshop's objectives are based on these aspects, bearing in mind the relations between subjects and environment, on which our capacity to acquire and use knowledge depends. In his article "Discipline and punish. The birth of the prison" Michel Foucault, the French historian and philosopher, explains how the layouts of the first school buildings dating back to the Eighteen-hundreds adjusted to an educational methodology resulting directly from military discipline and industrial logics. *"The organisation of a serial space was one of the great technical changes in the elementary school. It enabled the traditional system to be surpassed (a pupil who worked for a few minutes with the teacher while the disorderly group of those waiting remained idle). Given individual places, with the control of each individual made possible and the work simultaneous for everyone, it organised a new learning time economy; it made the school space function like a machine for learning but also for supervising, grading and rewarding."*[1]

* Fabrizio Pusceddu, DADU, University of Sassari.

[1]. Foucault M., *Sorvegliare e punire: la nascita della prigione*, Italian trans., by Alcesti Tarchetti, Einaudi Paperbacks, N° 77, then ET N° 146, Einaudi, 1976.

Nowadays, however, we are aware of how learning mechanisms are mostly connected with the capacity to act in a dynamic, selective way in space. According to Joan Stiles[2], the development of cognitive ability already begins at embryo level, from the twenty-eighth day after conception, and continues in a linear manner up until the fifth year of age. After that, rather than evolutionary genetics and development, we may speak more of an update based on the personal construction of experiences. It is common knowledge, apart from being scientifically proven, that cognitive capacities, "intelligence", do not depend on the size of the cerebral mass so much as the number of neuron links (synapses); at birth we know that a child possesses over 50 trillion synapses, which become 1,000 trillion at the age of 3, hence with extremely rapid development per unit of time, to then gradually decrease to 500 trillion at the age of 15. The cerebral functions relating to learning are not therefore "engraved" at birth like the capacity to react to other types of stimuli or functions (see the case of cardiac rhythm control), but are closely linked with the use made of them and therefore of the experiential components affecting the subject's life. In order to survive, the synapses need to be continuously stimulated and trained, and if we could have control over these operations, they would be done for precise areas, as particular cerebral areas are involved in different forms of learning (such as, for example, the area controlling language, which reaches full development between 6 and 12 years). From the point of view of space design, it means building potential opportunities to stimulate experience, *"the city instinctively senses all possible projects that link us interactively with its affordances, to use Gibson's term, i.e. with its "resources", the architectural, spatial, functional grips that it offers our capacities of perception, interpretation and action".*[3]

Learning is no longer, therefore, solely a cognitive fact where all the components of the context are present in a passive form aiming at the maximum control of concentration and attention, but a condition directly linked with the need to act. To act we need knowledge, but cannot have access to it except by acting. It is a simultaneous, reciprocal process that besets our need and capacity to select and leads us to exclude parts of the world to concentrate on the aspects and elements that are mostly of interest for our purpose. Berthoz identifies the reasons for this process in human beings' need to reciprocally link the perceptive component with the motor component[4], the "perceiving something in terms of" and the "doing something in terms of". At the centre of these operations is the brain, which becomes a machine of convergence between the two components, as well as being a developer of information and simulator of actions.

---

[2]. Eberhard J.P., *Brain Landscape – the coexistence of neuroscience and architecture*, Oxford University Press, 2008, p. 48.

[3]. Tagliagambe S., *The Territorial Future of the City – the dilation of the concept of inhabit and the city/territory relationship*, Springer, 2009

[4]. Berthoz A., *Le sens du movement*, Odile Jacob, Paris, 1997, Italian trans. Il senso del movimento by E. Dal ra and A. Rodighiero, McGraw-Hill, 1998. "The brain is a simulator (...) This means that it is the whole of the action that is simulated in the brain by internal models of physical reality that are not mathematical operations but true neurons whose properties of form, resistance, oscillation and amplification are part of the physical world and tune up with the external world. Which suggests that the brain deals with movement following two methods. The first is conservative and functions as a subjugated system, while the other, a projective one, simulates the movement to foresee its consequences and choose the best strategy."

Supported by the most recent scientific discoveries[5], this responds to Walter Benjamin's doubts: *"We may be more or less used to the gesture of grasping a cigarette lighter or spoon, but we know practically nothing about what actually takes place between the hand and the metal, not to mention the way in which this varies in relation to the state of mind in which we find ourselves"*[6], thus understanding *"the distracted perception of the mass as the only true knowledge of architectural work.*[7]
Architecture, in contrast with other arts, does not require attention to be read and understood, but needs to be experienced with distraction and automatism, via a process of spontaneous appropriation of the space by subjects. Architecture cannot be imagined as detached from its context, but above all *"architecture cannot be an object that can only be observed visually"*.[8] The perception of a space takes time and consideration, it means living within it in time that cannot be instantaneous, interacting with it, grasping its dimensions and regulating one's own movements with respect to them, and all this in a *distracted manner,* automatically, spontaneously at different levels of attention or organising structures. Design for learning space, or more specifically for the school, cannot avoid these mechanisms or neglect to consider subjects as active elements in the space and not simply receptors of external stimuli. If the perception of space is *distracted,* the design of it is not. In order that people may guide their own personal projects of knowledge and action in the world it is necessary that the world provide reference points, guidance that will not turn the distraction, automatism and spontaneity of behaviours into bewilderment over the loss, excess, confusion or mixing of clear elements of reading and interpretation. Space as an engine of development of processes and actions takes a revelatory role towards its surroundings and suggests in a spontaneous manner what its future evolution might be. And if it is true that the heart of the learning process is encompassed in the act of perception, be it aware or unaware, then in the relations space is able to generate between action and information, in particular in school environments, we can identify a clarifying contribution to solve our initial misunderstanding. Schools not as containers of knowledge to tap into in a reflective way, but active spaces stimulating experience.
From this background and to follow on from the various aspects of the interventions on the school building patrimony, without deviating from the real problems but, on the contrary, trying to anchor opportunities for intervention to them, the case studies were proposed in the workshop. The choice for singling out project spheres focused for reasons of proximity on the territory of north-west Sardinia, between the towns of Alghero, Sassari and Porto Torres, strategic contexts for drawing up design proposals at different scales.

---

5. Rizzolatti G., Sinigaglia C., *So quel che fai*, Milan, Cortina, 2006. Action is not a pure consequence of a process of reading and interpretation, as was thought for a long time, but a moment of understanding of a state of things in the world, where perception becomes an integral and indiscernible part of it; it is a process that is not structured in clearly separate phases, but in the achievement, actual or simulated, of "motor acts" that lead to defining the behaviour of subjects not as "simple movements".
6. Benjamin W., *Das Kunstwerk im Zeitalter seiner technischen Reproduzierbarkeit*, Suhrkamp Verlag, Frankfurt, 1955. Italian trans. by Enrico Filippini, *L'opera d'arte nell'epoca della sua riproducibilità tecnica*, Einaudi, III Ed 2000, p. 43.
7. Ibidem, p. 44 – 46.
8. Ibidem.

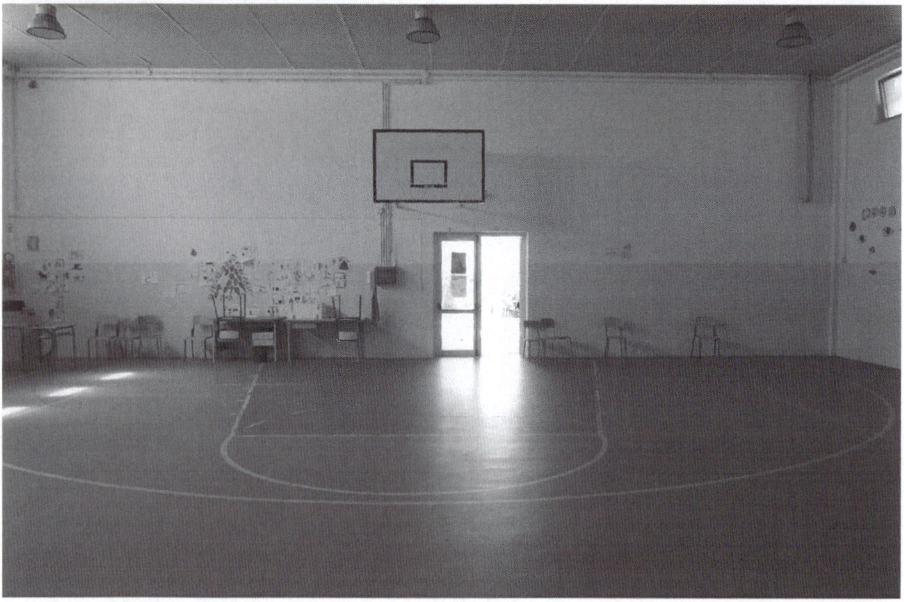

Some spaces of the school

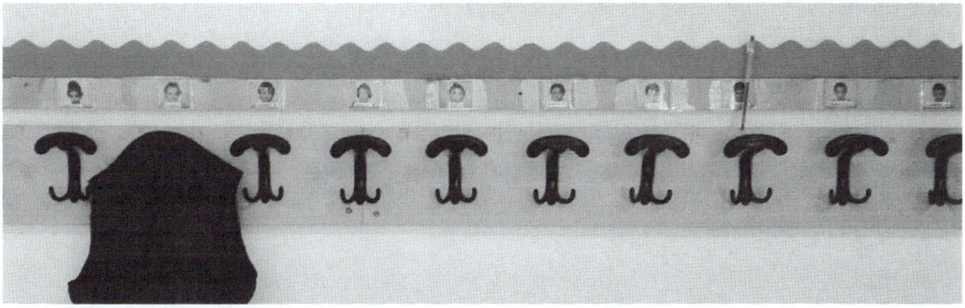

5
WORKSHOP

ILS project workshops

5.1

AH01

The School of "Sacro Cuore"

Gianluigi Mondaini with Lino Cabras and Andrea Dessì, Elena Melis, Sara Mozzo, Roberta Scarpa, Matteo Sotgiu, Barbara Zaru

Type: Kindergarten and Primary School
Realization: 1903
State: in use
Focus: the school building, rigorous in its formal and decorative aspects, is located in front of the "Giuseppe Manno" public gardens with which it has no material or immaterial connection.
Project: intervention on the existing
Municipality of Alghero

The school of "Sacro Cuore" as a new learning landscape
Lino Cabras

The institute of "Sacro Cuore", set up as primary school in 1903, is now part of the comprehensive institute of Alghero. After more than one century from its construction, the building is now considered as one of the most remarkable monuments of the city, example of classic architecture characterized by a symmetric horseshoe structure. The inner space, austere and severe, is set on the strict opposition between corridor and classroom.
The giant scale of the building, both in plan and elevation, such as the lack of shared spaces for socialization, make really difficult to conceive the contemporary didactic, mainly based on informal and hybrid spaces.
Undermining the space structure of the school, in its inner and outer space relations, becomes the main goal of the project.

The denied relations with the surroundings (the historical city within the walls and the public gardens facing the main elevation of the building) are retrieved working on different gradients of interactions between the school and the city activities.
A new shape of the soil between the school and the public gardens defines a set of new playgrounds, conceived as protected courtyards and placed one level below the public space. On the level of the city every physical limit is avoided, creating a unique common space.
A new facade under the current level of the street faces the public gardens with an hypogeum architecture hosting inclusive activities for the city, a social center and artistic workshops.
On the rear facade, the project mounts a system of urban connections between the commercial square and the new suspended building of the gym, open to the city. Under the floating volume a covered square is dedicated to the commercial activities in connection with the old marketplace of Alghero.
The project combines all the urban functions based on the "exchange" in its widest meaning and puts together different urban fragments, currently stuck in a contemplative condition.
The interior space aims to create a new relation between the learning spaces and the connection areas by inserting a "parasitic" wooden element conceived as a spatial device growing from the floor and walls of the historical building to define an inner landscape able to overcome the preexisting partitions, treated with cuts and transparent inserts.
Minimal intervention and flexibility are the main criteria used in the project. Continuous change of points of view are created to enable children to explore the hybrid space among the classrooms, while the corridors become a unique learning place where the didactic functions are de-spatialized.

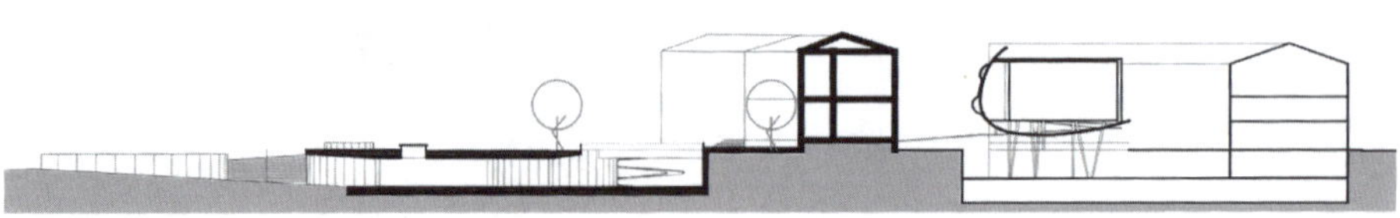

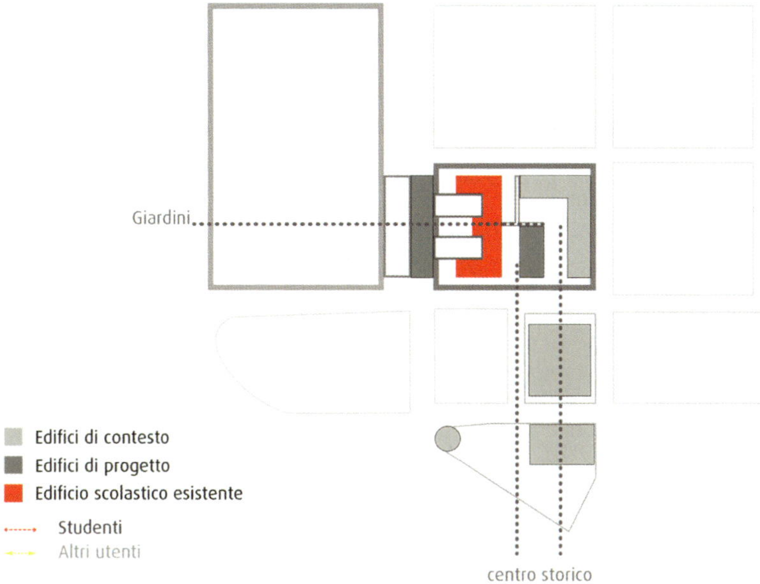
Giardini
centro storico
Edifici di contesto
Edifici di progetto
Edificio scolastico esistente
Studenti
Altri utenti

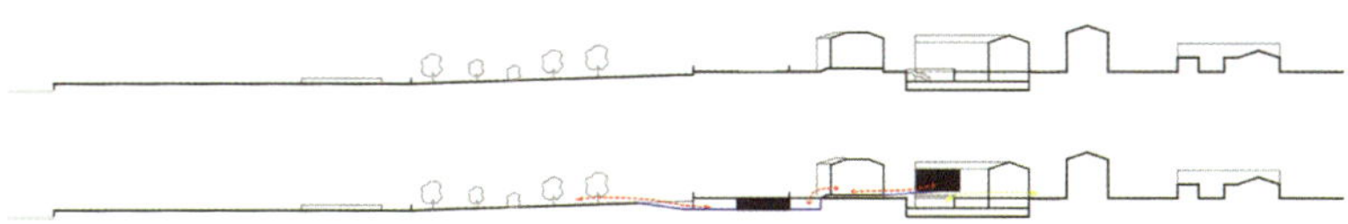

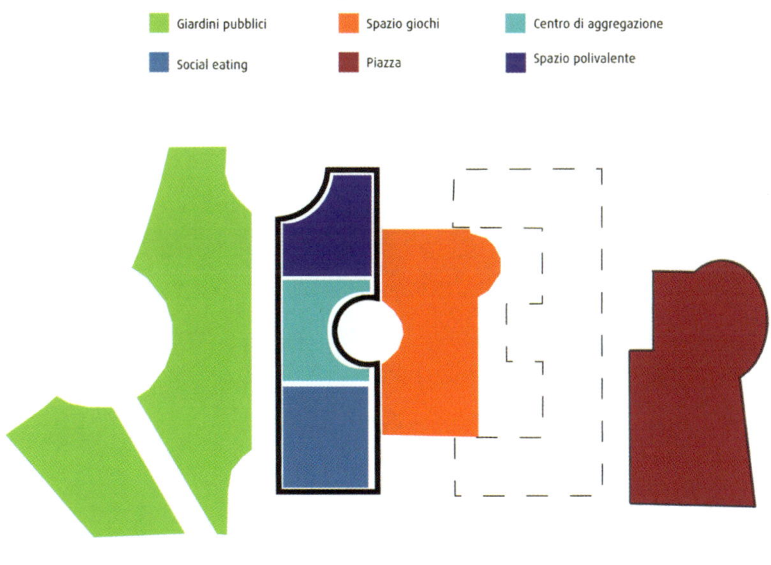
Giardini pubblici
Spazio giochi
Centro di aggregazione
Social eating
Piazza
Spazio polivalente

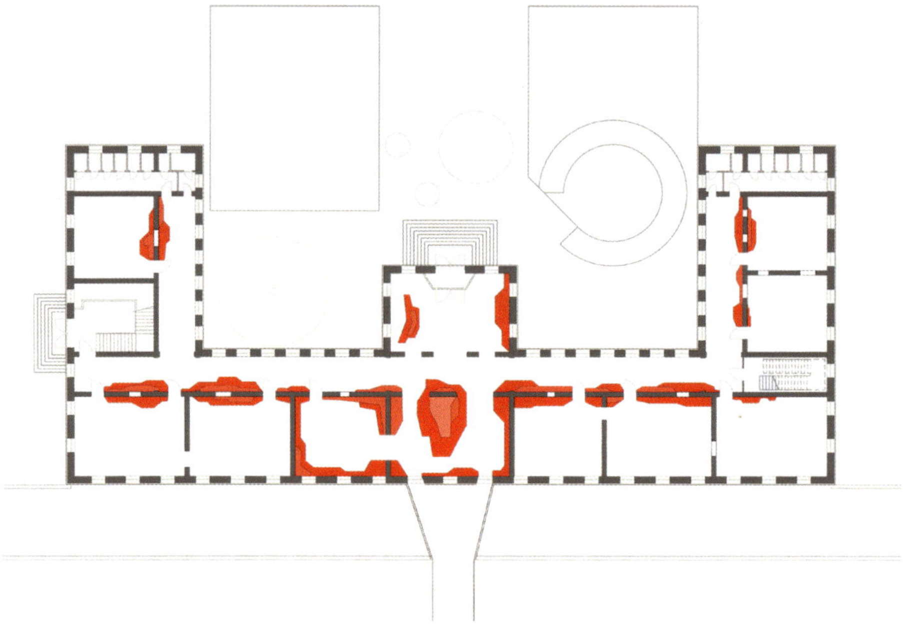

5.2

AH02

New micro-crèche for "La Pedrera" Unified School N° 2

5+1AA Alfonso Femia, Simonetta Cenci with Samanta Bartocci, Fabrizio Pusceddu and Stefania Cabras, Stefano Delogu, Andrea Lai, Benedetta Tilocca, Chiara Tilocca, Marco Tradori

Type: Micro-crèche for very young children
Focus: the Unified School N° 2 of Alghero came into being in 2012 combining several schools at different levels – Infant, Primary and Secondary. The idea was to create a new facility aimed at accommodating very small children (0-3 years).
Project: new building
Municipality of Alghero

The metamorphosis of a place: new micro-crèche for "La Pedrera" Unified School

Samanta Bartocci, Fabrizio Pusceddu

The tale of the *"possible metamorphosis of a place, between dream and reality, between beauty and responsibility"*. This is how Alfonso Femia and Simonetta Cenci, steering the working group to create the new micro-crèche at the Pietraia, Alghero, describe their project. For basically, the context is still that of a suburb where the founding order of agricultural mesh had to make room for urban requirements that, over time, did not manage to generate urbanity but solely fragments of buildings devoid of any relational identity. But it is through design that that we can remove those layers lacking in sense that have gradually made the meaning of the places invisible, or almost, and marginal: the agricultural, cultural and urban landscapes. Efforts are being made to anchor the school to its surroundings, re-reading the relationships between full and empty elements and defining a new hierarchy of limits and borders. It is first and foremost the attribution of a role, the opening up of an area by declaring its perimeter, marking points of access and entrusting one of these with the strength to determine an axis of permeability across the whole plot, which will be able to plastically change the existing school building, breaking up its volume and creating a small internal courtyard at the main entrance. Attention to the lay of the buildings and their aspect within the project area then does the rest, dictating the hierarchies between the different perspectives and suggesting the ideal spot for the new nursery space. Design is therefore like a toolbox (boîte à outils) between real and imaginary, which, with its openness, gives back work categories for the space that are clear and controllable:

1. the surroundings, as a redefinition of a sequence of increasingly open enclosures in school-city relations;
2. the pathway, as a system to foster relations, the meetings and flows within a single space, between the parties. The enclosure and the pathway define a new geometry which reveals, in the weaving of the ancient olive groves, a sequence of spaces each with their own qualities and features, in terms of both use and materials. The volume of the crèche is an integral part of this and a generating element that directs the routes and points of access towards the existing elementary school. The green area, duly structured in the weaving of the trees, is the element of measure of the space, able to "colonise" the external urban space. Whereas the parking system finds room in the bending of the external limits of the block, while the play areas cover the space close to the buildings in question. A suspended terrace over the internal courtyard creates a perceptual device over the new landscape of the children, while the need to protect the façade to the south from direct sun's rays becomes an opportunity to reconsider the urban front in a dialogue with the public space.

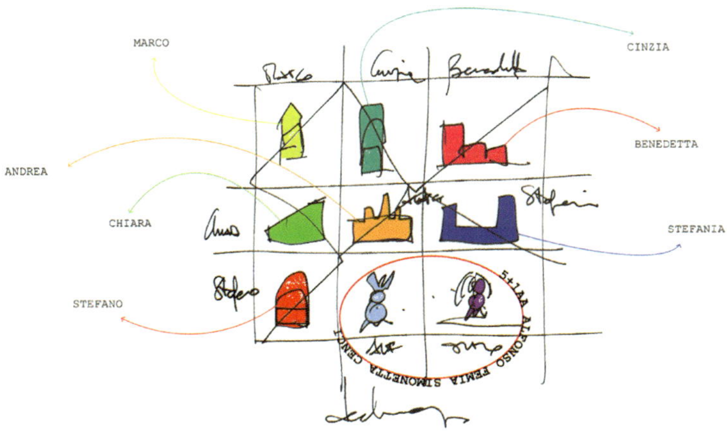

VI RACCONTANO
LA POSSIBILE METAMORFOSI DI UN LUOGO,
TRA SOGNO E REALTÀ,
TRA BELLEZZA E RESPONSABILITÀ:
IL NIDO DEL ~~CONTE~~.PETRAIA

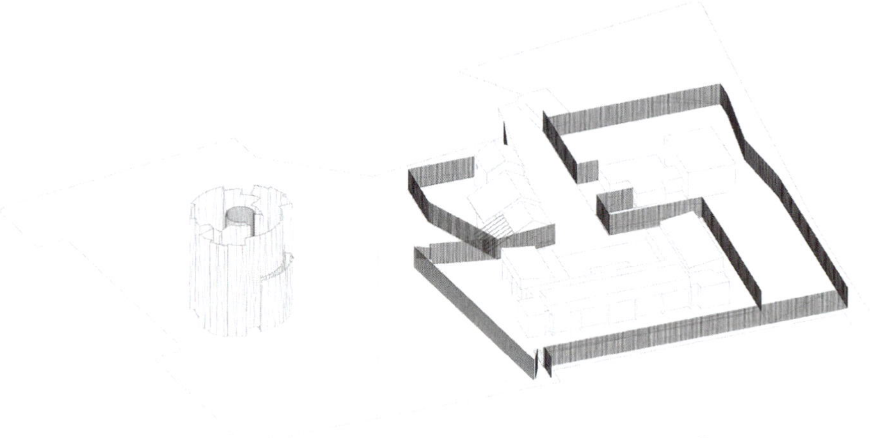

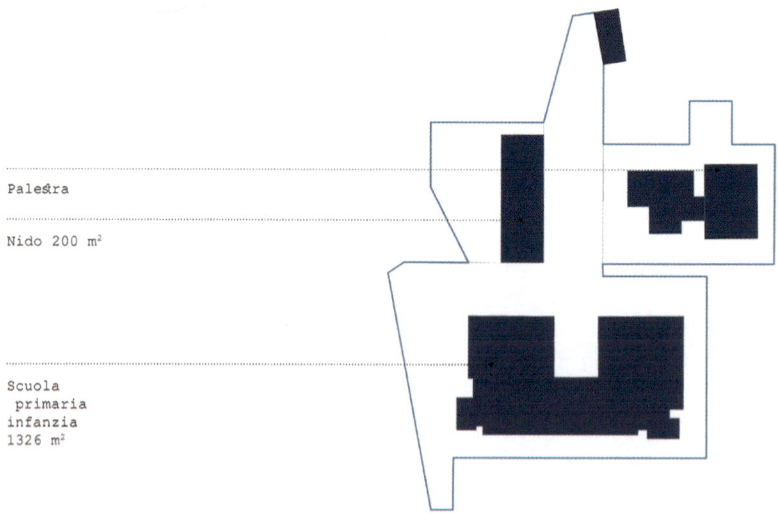
Palestra
Nido 200 m²
Scuola primaria infanzia 1326 m²

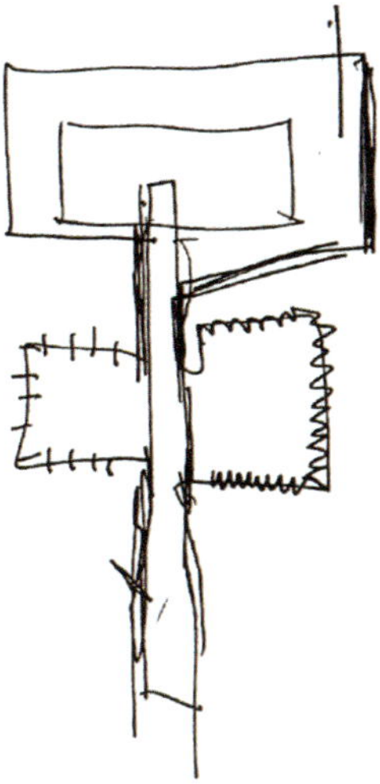

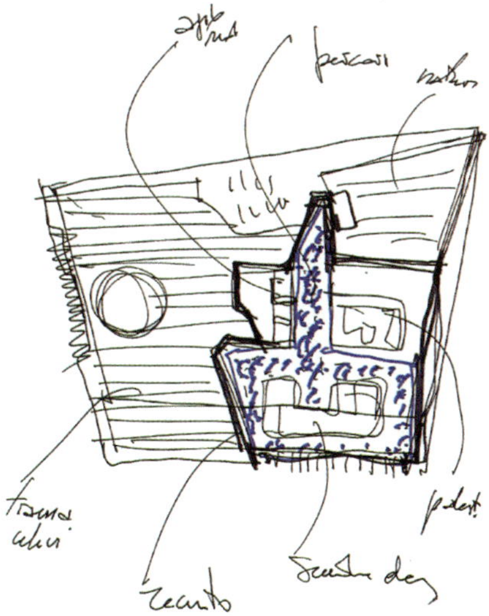

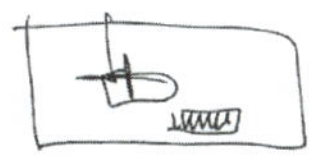

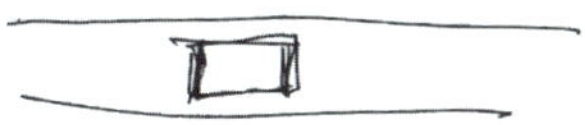

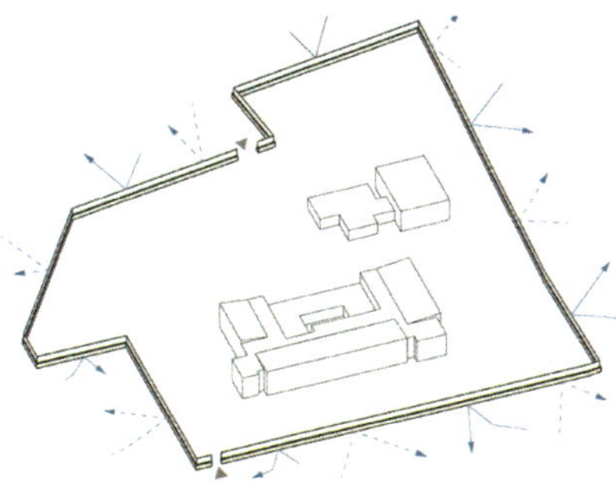

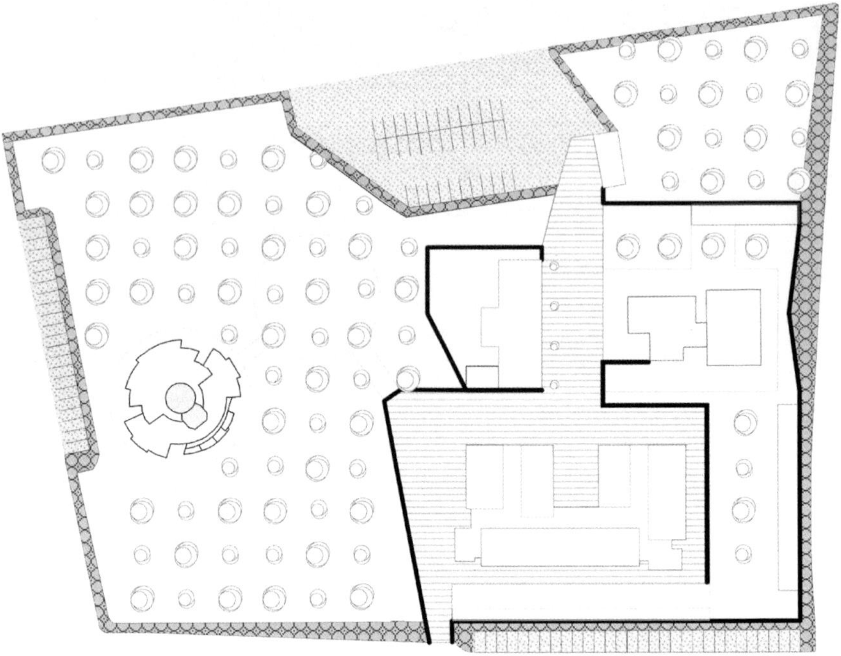

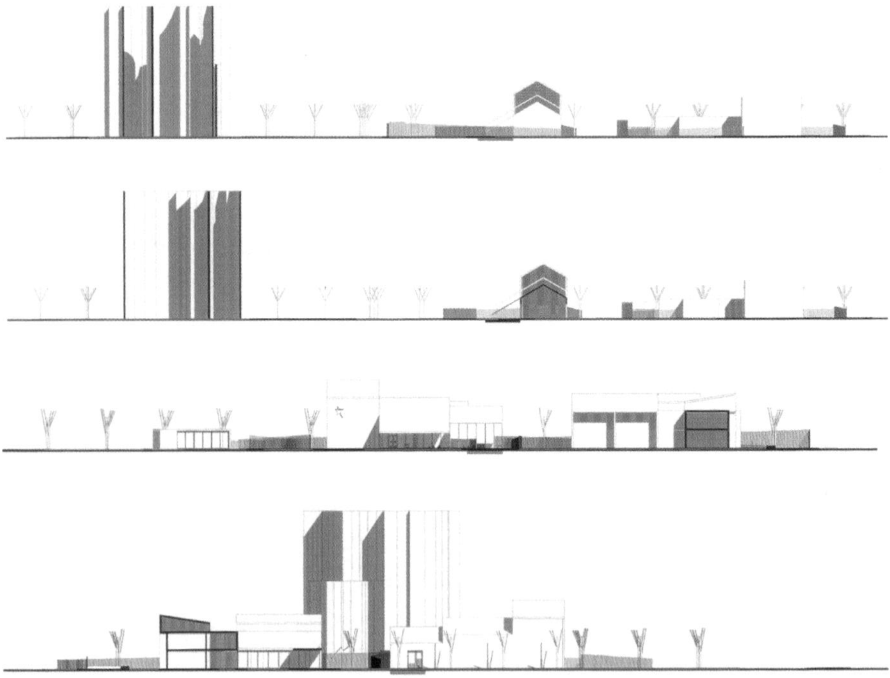

5.3

SS3

Extension of State Middle School N° 11

Massimo Ferrari with Claudia Tinazzi and Gian Mario Dettori, Marco Matia Fiori, Michele Lucesoli, Alighiero Lucidi, Davide Perrotta, Marco Rosciani

Type: Level 1 Secondary School
Built: 2007
Condition: in use
Focus: the school accepts students from a vast territory comprising the hamlets of Li Punti, San Giovanni and Ottava, also including the country areas of Pian de Sorres, Villa Gorizia, Viziliu and Truncu Reale. The expansion of the town of Sassari towards the countryside requires greater provision of school spaces and the consequent extension of school complex N° 11.
Project: extension
Municipality of Sassari

Just listen to the walls
Claudia Tinazzi

"Every work speaks more freely, if their author is unknown".
Maria Lai

An architecture project is often just a pretext, an occasion, a ruse to verify the honesty of a well-founded reflection. Even better, today it represents the best opportunity to try to make a reasoning reach far, like when we work full time on a certain theme, devoting our study and research to it, but without ever actualizing our ideas. No preconception or prejudice is allowed, when starting a new project When faced with a contingent reality, the critique of what's been done before, the points of reference we chose, and the principles we followed to that point are rearranged to form new hierarchies, where unforeseen connections and solutions of continuity reveal unexpected reasons.
The project for the enlargement of the building that houses the middle school "Li Punti", in Sassari,

was developed during the ILS (Innovative Learning Spaces) workshop and offered the opportunity to work free from any predetermined idea. Designed in 2007, the school building is inadequate from the functional, spatial and architectural point of view, and for this reason, our first idea was to demolish it, rather than to enlarge it. However, after discussing with several educators, who draw our attention on the fact that a radical change of the teaching environment would have a disruptive effect for those who frequent it every day, we decided to accept the challenge and to engage in hand-in-hand combat with the existing building.

The change was driven by walls: re-encountered in the memory of places, and providing solid base for many well-known projects, like the houses designed by Alberto Ponis and, more recently, by Umberto Riva. These contemporary borders have always been present on the island and reflect the building skills of the past. During the 20th century, these walls - in their simplicity - have been used to build spaces for teaching, learning, and playing, like in the works by the BBPR group or by Aldo van Eyck. Walls have been used to rearrange spaces and opportunities, and the existing structure survives in the re-designed spaces like a ghost, or like the skeleton of the new building.

Intermediate education requires the presence of classrooms, but this traditional space has been rethought with sincerity and the principle of adjacent sections has been implemented. A series of spaces – divided by partitions that are intentionally fragile and that are independent from the wall line yet follow it – provides a new interpretation of the relationship between the inside and the outside. Inside the school, the fixity of distribution spaces has been overturned in order to allow for a necessary dialogue among the different sections. Without knowing it, the project patiently followed the existing geometry, and enhanced through simplicity the characteristics of a place with its anthropological peculiarities.

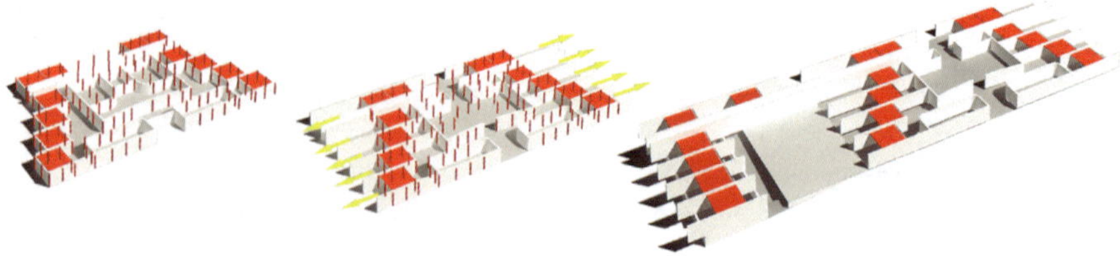

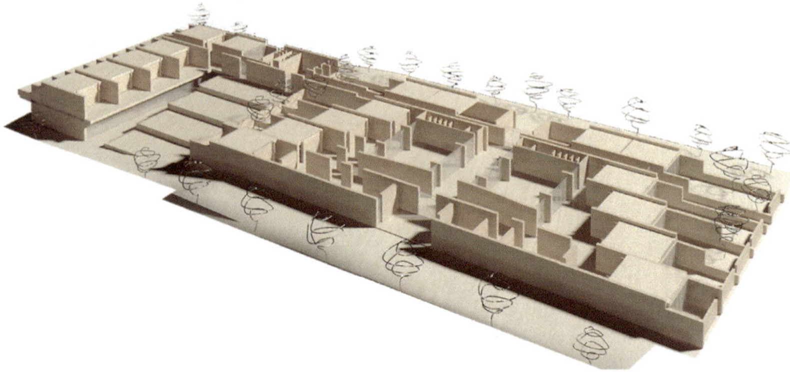

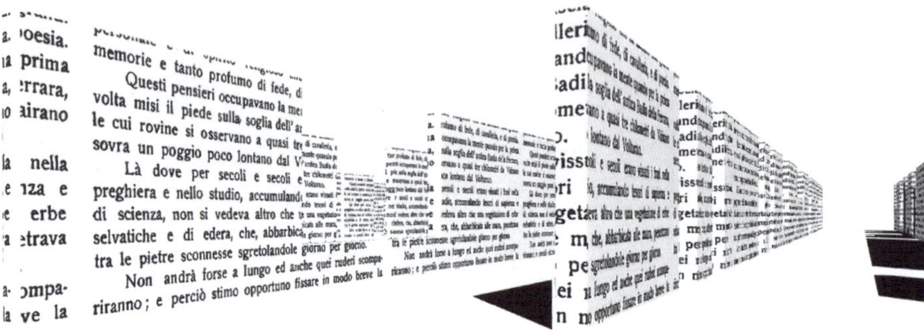
memorie e tanto profumo di fede, di
Questi pensieri occupavano la me
volta misi il piede sulla soglia dell' a
le cui rovine si osservano a quasi tr
sovra un poggio poco lontano dal V
Là dove per secoli e secoli e
preghiera e nello studio, accumuland
di scienza, non si vedeva altro che
selvatiche e di edera, che, abbarbica
tra le pietre sconnesse sgretolandole giorno per giorno.
Non andrà forse a lungo ed anche quei ruderi scompa-
riranno; e perciò stimo opportuno fissare in modo breve la

5.4

SS4

Elementary School N°11 and the residence for the elderly

Matteo Scagnol, Sandy Attia with Francesca Arras and Eleonora Fanesi, Giorgio Marongiu, Claudia Pintor, Andrea Scalas, Serena Valeri

Type: Level 1 Secondary School
Built: in the 1950s
Condition: in use
Focus: the building consists of an original block and subsequent extension (of the Nineties). Apart from special maintenance, it needs a general reappraisal of the space and connections with the surrounding area, particularly its relations with the nearby valley, the facility for the elderly and the rest of the town.
Project: conversion
Municipality of Sassari

The school as an enabling factor and device[1]
Francesca Arras

Working with the group of participants, as diversified[2] as stimulating, led by Modus Architects (Sandy Attia, Matteo Scagnol), proved an interesting opportunity to test some ideas on what it means to plan truly *innovative* educational spaces, in relation to the specific requests and desires, that emerged from the ongoing dialogue with the Via Gorizia School of Sassari, that were felt could be granted immediately. The experiment explored the possibilities of rethinking the school setting as a trigger for the design of an instrument that could really *enable* and connect with the city through the redesign of its general configuration and its spatial relationships. We developed three proposals with different feasibility gradients, organized by phases and scales of intervention.

The common choices for all subgroups:

- start a gradual restitution process of the streets around the school by introducing social and relational uses. Reverse the current condition of the space, primarily designed for transit and the parking of motorized vehicles, in favor of different forms of cohabitation between modes of motorization and pedestrians, as well as for collective practices (playing, gatherings, exchanges, etc.). Address the block redesigning issue with the creation of a new macro-block, promoting walkability, and merging together the current kindergarten, primary and middle schools, and the seniors residence "Casa Serena", as well as reconnecting this new entity with the system of valleys which it faces;
- disengage the current point of access from via Gorizia, creating a large public space, thus solving the school's access issue and promoting cross-permeability from the main roads into the valley and urban interaction. Working with calibrated additions and subtractions, demolition, elevations, landfills and terrace construction, to consolidate a solid relationship between the new block system and the valleys
- producing an area of progressive territorial cohesion and providing services to the community, thus becoming a true civic center and community hub;
- revolutionize the interior spaces and the concept of traditional classrooms in favor of classrooms sectioned into a cluster system with learning landscapes, to promote relational-laboratorial learning activities, through the conceptual innovation of the interior architectural spaces and their various functionings.

References

- Arras F., Ghisu E., Idini P., Talu V. (2013b), *'TaMaLaCà - Tutta Mia La Città. 'Suffragette' dei diritti urbani negati'*, Atti della XVI Conferenza Nazionale SIU *Urbanistica per una diversa crescita*, Napoli
- Hertzberger H. (2008), *Space and Learning*, 010 Publischers, Rotterdam
- Lefaivre L., De Roode I., (2002), *Aldo van Eyck. Playgrounds*, NAi Publishers, Rotterdam
- McLuhan M., Hutchon K., McLuhan E., (1977), *City As Classroom: Understanding Language and Media*, Book Society of Canada, Agincourt, Ontario, trad it. Lorenzini A. (a cura di), (1984), *La città come aula. Per capire il linguaggio e i media*, Armando Editore, Roma
- Talu V. (2014), *Qualità della vita urbana e approccio delle capacità. Perché e come promuovere le capacità urbane degli abitanti più svantaggati*, Franco Angeli, Milano
- Weyland B., Attia S., (2015), *Progettare scuole. Tra pedagogia e architettura*, Guerini, Milano
- Ward C. (1977), *The child in the city*, Architectural Press, Princeton, trad it. (2000) *Il bambino e la città. Crescere in ambiente urbano*, L'ancora del Mediterraneo, Napoli

[1]. This essay is closely linked, for setting and bibliographic references, to the paper "The School as a Public Space [and the City as a Classroom]" led by Tamalacà.

[2]. E. Fanesi, S. Valeri, engineer graduates of The Polytechnic University of the Marche; C. Pintor, G. Marongiu, A. Scalas, recent architectural graduates of DICAAR; M. F. Ghiaccio, pedagogist and research fellow with DISSUF-UNISS, we must also include Barbara Letteri e Donatella Olmeo, teachers of the VI C.D. school of Sassari.

BIBLIOTECA
AULA
AULA
AULA MAGNA
AULA
MENSA
PALESTRA
ACCESSO
AULA
AULA
AULA
AULA
GIARDINO
AMMINISTRAZIONE

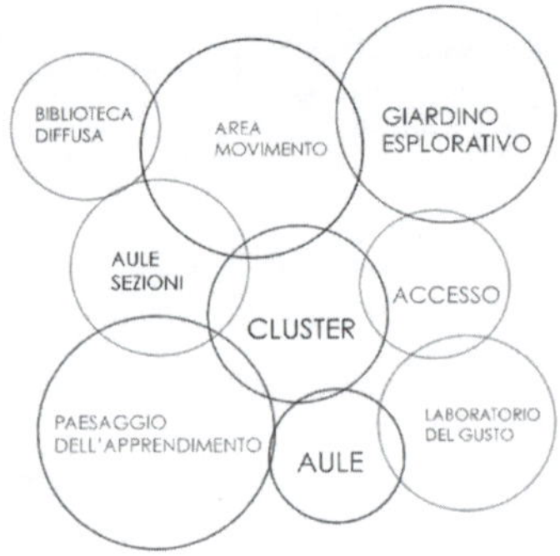
BIBLIOTECA DIFFUSA
AREA MOVIMENTO
GIARDINO ESPLORATIVO
AULE SEZIONI
CLUSTER
ACCESSO
PAESAGGIO DELL'APPRENDIMENTO
AULE
LABORATORIO DEL GUSTO

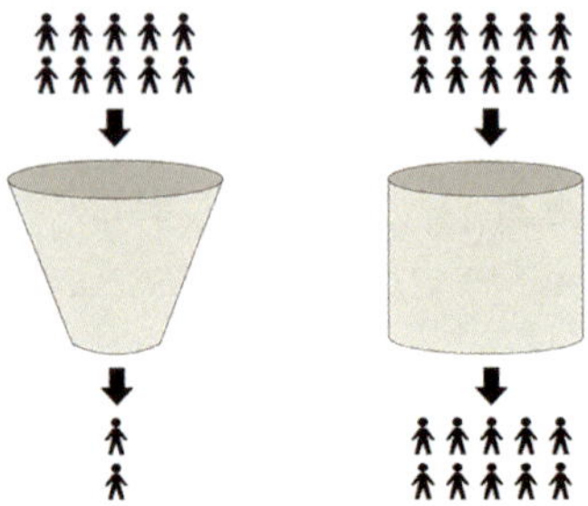

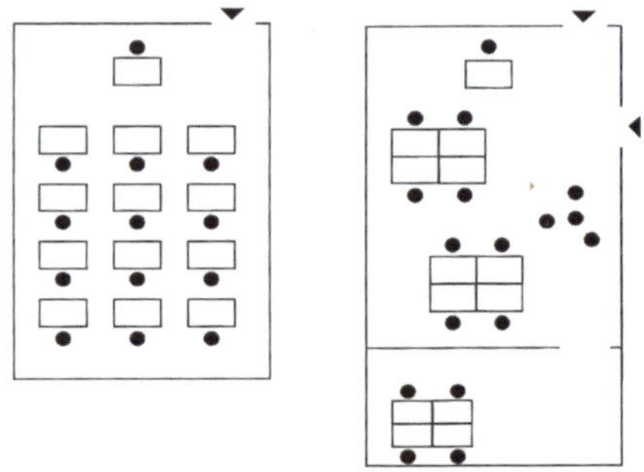

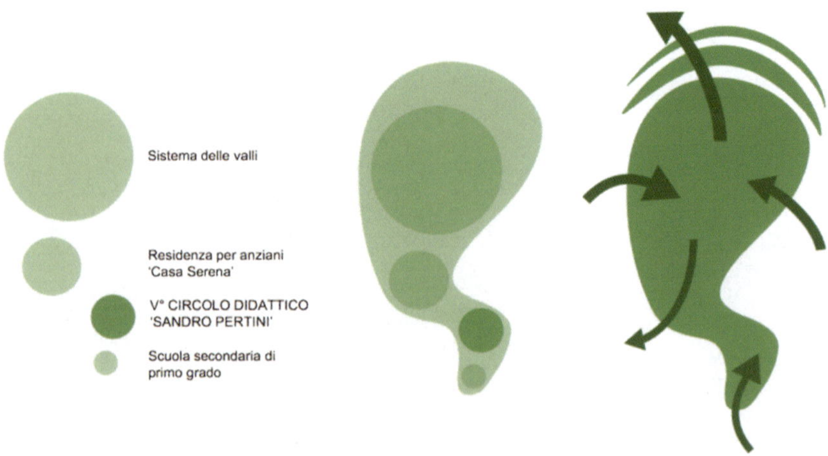
Sistema delle valli
Residenza per anziani 'Casa Serena'
V° CIRCOLO DIDATTICO 'SANDRO PERTINI'
Scuola secondaria di primo grado

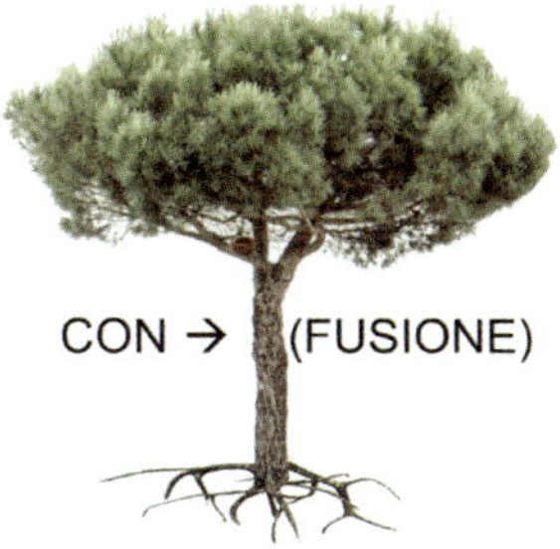

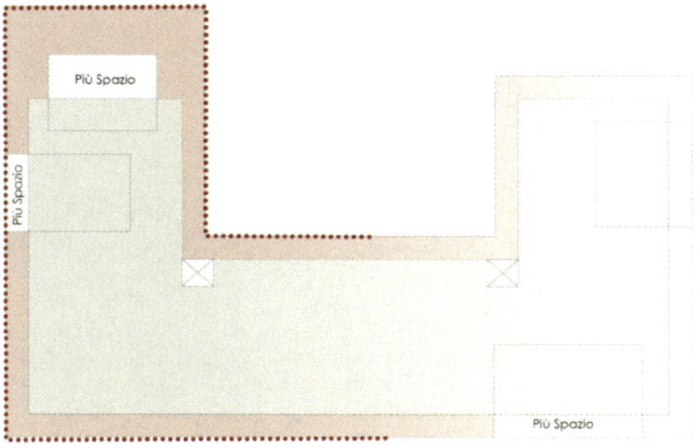

DEMOLIZIONI

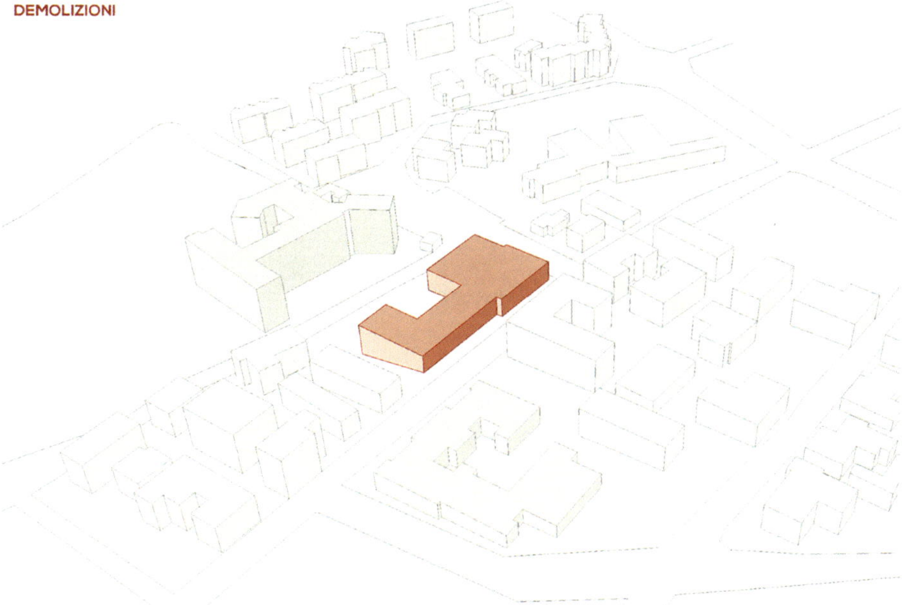

PIASTRA - connettere orizzontalmente

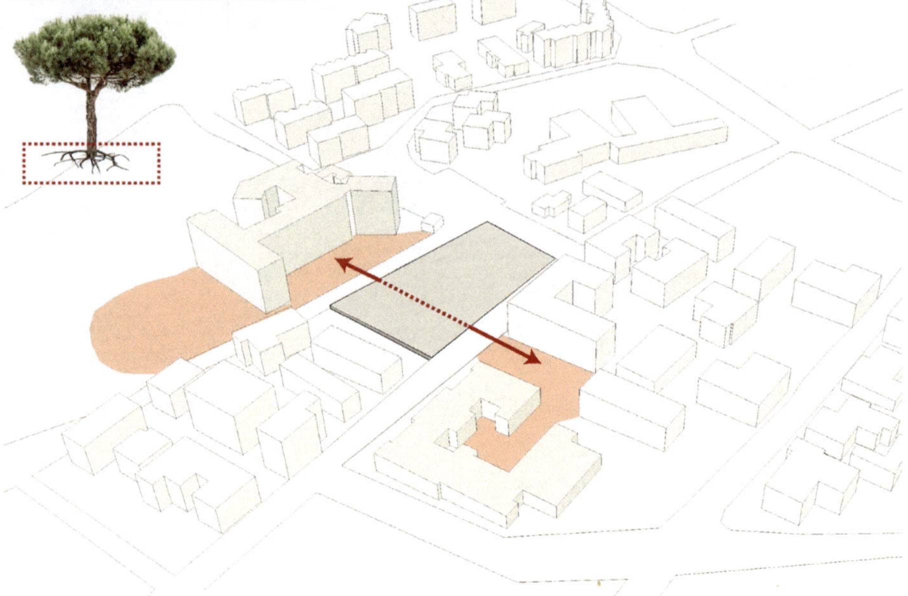

NIDO

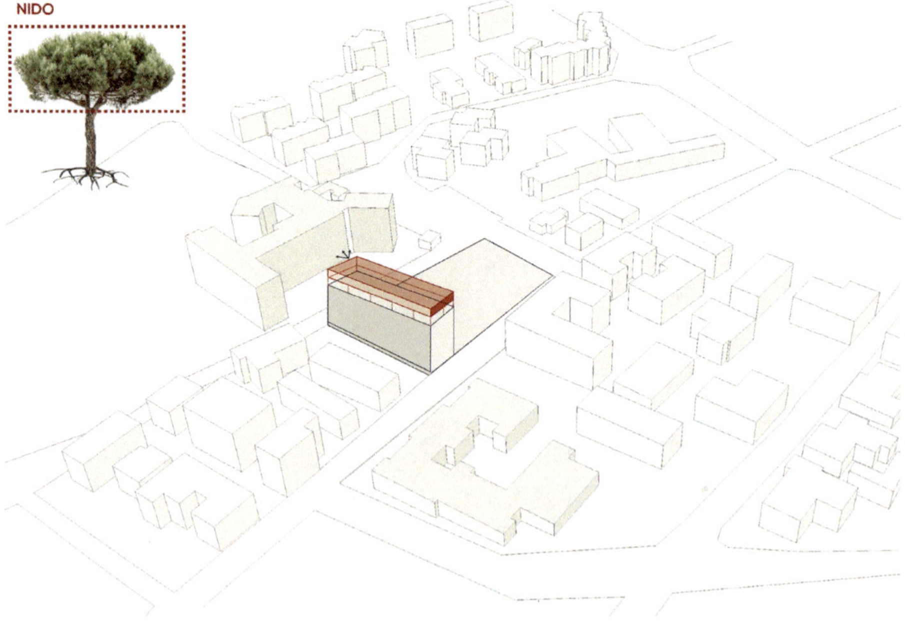

BOSCO

5.5

PT5

Ex Elementary School De Amicis

Patrick Fransen with Francesca Rango and Simone Antoniucci, Debora Benfatto, Alessandro Bisciardi, Alessandro Fenu, Ludovica Giaconi, Marco Zaccheddu

Type: Primary School
Built: 1912
Condition: partly in disuse
Focus: no longer in use for the last few years, the school is the focus of attention of some citizens' associations that would like to keep up the past memories and original function of the building no longer suitable today for scholastic use. "We consider it right that the authorities intervene immediately to stop its deterioration, as we should not forget that this school represents the past memories of several generations who obtained their first notions of studying and awareness there."
Project: conversion
Municipality of Sassari

The school as a model of relationship
Francesca Rango

Most of the words of the semantic sphere of "learning" suggests a certain type of relationship between the subject and the object. Over time psychology, pedagogy and didactics have declined those relationships in different paradigms in order to answer the questions "How do you know?" "Which is the nature of the mind and how is it organized? "," How teaching (external action to the subject), can favour the learning (internal action to the subject)?". A question, the latter, which invokes an active role of the learner through the word *Learning*, while it moves the action on the teacher through the word *Teaching*.

Beyond all expectations, the word *School* - the one most commonly used to indicate an environment dedicated to learn / teach- does not suggest any hierarchy between the subject and the object. The etymology of "school" comes from the Greek word *skholé* idleness, taking time, resting, ease and comfort.

On one hand, the words let us think of different relationships between subject, object and action. On the other hand, the spatial articulation of a school building considered as a social object (Markus, 1993) suggests the model of relationship between the subjects that inhabit/undergo the interior spaces and the relations between the set of these and the external environment. Overwhelmingly since the Enlightenment, these relationships have become an object of study for spatial and typological models to make the transmission of the *discipline* the most efficient, also through the control of bodies in the space (Foucault, 2005). From the panoptic education models of Bentham to the proposals of the utopian socialists, the underlying concept is that individuals can be "formed" in a passive relationship between the subject and object.

The spatial scheme of this conception is present in many still in use school buildings. It can be easily recognized in the old classroom model based on frontal lectures and in the total closure of the building towards the surrounding urban environment (Fig. 1).

In the project for the school De Amicis in Porto Torres, a building from the early 1900s (fig.2), the students were invited to reflect on the relationships between the subjects that spaces have produced and that can produce.

Not always the old buildings are the best suited to accommodate the new expectations. However, in some cases, there are the conditions to be taken in charge that go beyond the functional characteristics of the building. There are cases in which the building has a value for the collective imaginary, it is a landmark in the urban landscape; it is the holder of stories, even "minor", linked to the territory. The project for the De Amicis takes into account this significance, but it should not be considered only as a redevelopment project of a building, but a project for a kindergarten and primary, within an existing school building.

The design involves both the interior of the building in its school functions and the relationship with the urban surroundings. The guiding principles are: possibility to accommodate more functions, opening the building on several fronts, ground floor open to public functions, differentiation of learning spaces.

In particular, the layout of the latter is based on the assumption that learning can take place according to different modalities (fig.3). The space itself is an exhortation to different uses depending on the needs and rhythms of the single and the groups, a concept, which, although if not so close to a *skhole* meant as a space for idleness, at least brings back the meaning of ease and comfort and their different declinations.

References

- Markus T.A. (1993), *Building and Power*, Routledge, London
- Foucault M. (2005), *Sicurezza, territorio, popolazione. Corso al Collège de France (1977-1978)*, Feltrinelli , Milano

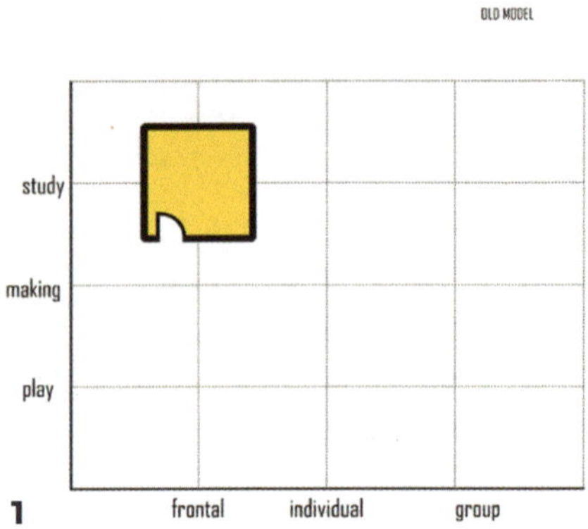

1

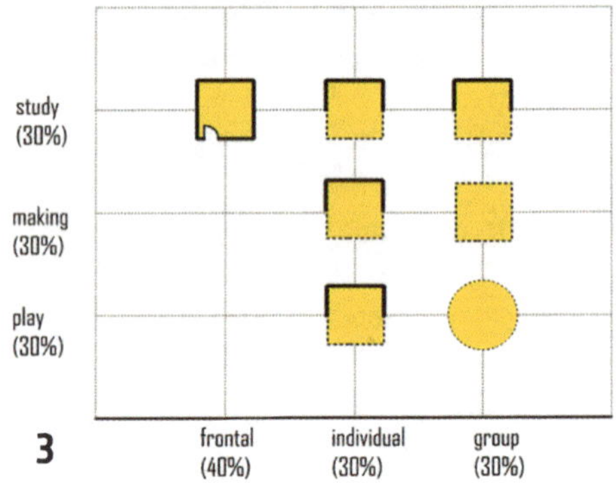

3

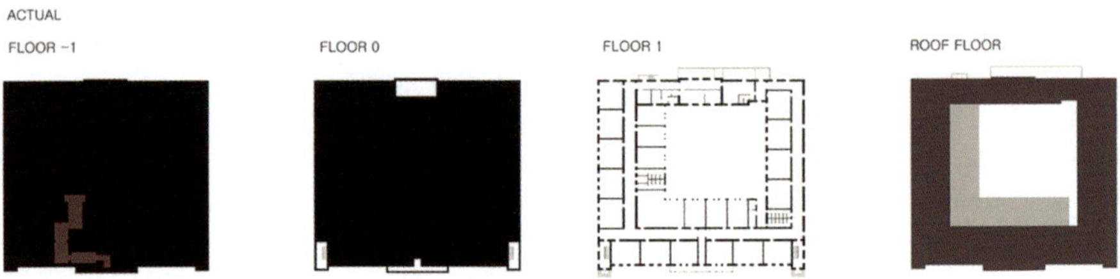

2

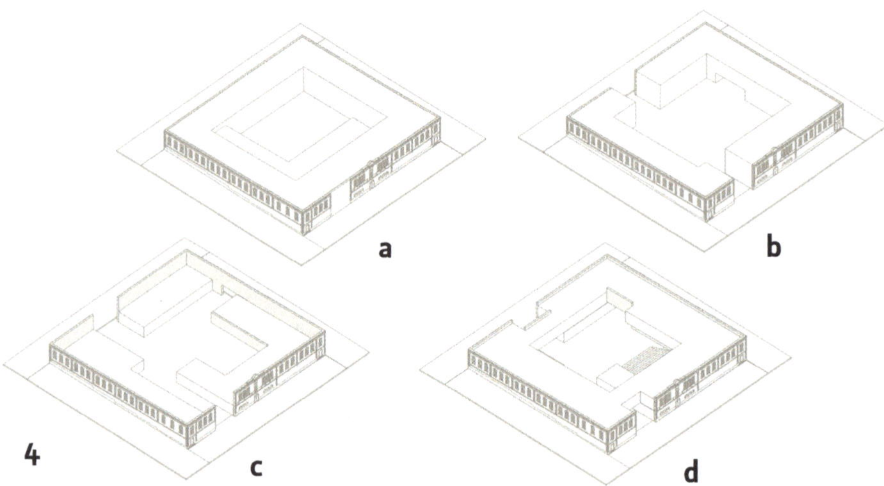

4

Fig. 1- In the predominant model the relationship between the organization of the activity (frontal lectures) and the action through which you learn (study) is based on a closed and mono-directional type of space.
Fig. 2- Istituto De Amicis, Porto Torres Current Situation
Fig. 3 – From the intertwining between the organization of the activity (lectures, individual or group work) and the actions through which you learn (to study, do, and play) is based on different types of spaces (open, closed, semi-open, though flexible)
Fig.4 - Current state of the building (a); New permeability (b); Ground floor and collective facilities (c); Kindergarten and primary school (d)

Fig. 5 - The school stands in relation with the urban surroundings activating public functions at the ground level.
Fig. 6 – Program
Fig. 7 – First level planimetric scheme, in detail the classroom of the primary school
Fig. 8 – Primary school, section
Fig. 9 – Building section

6

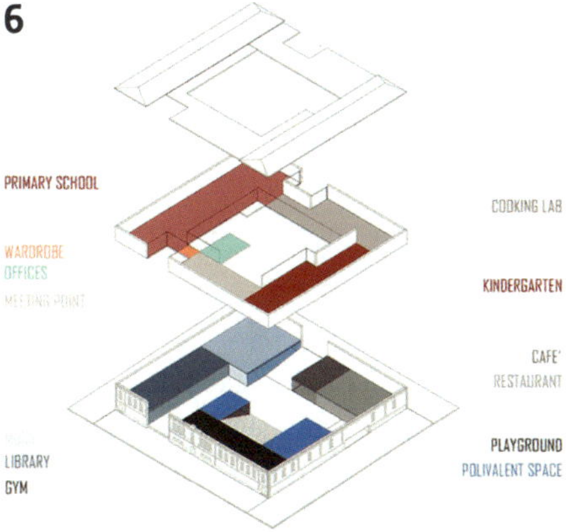

5

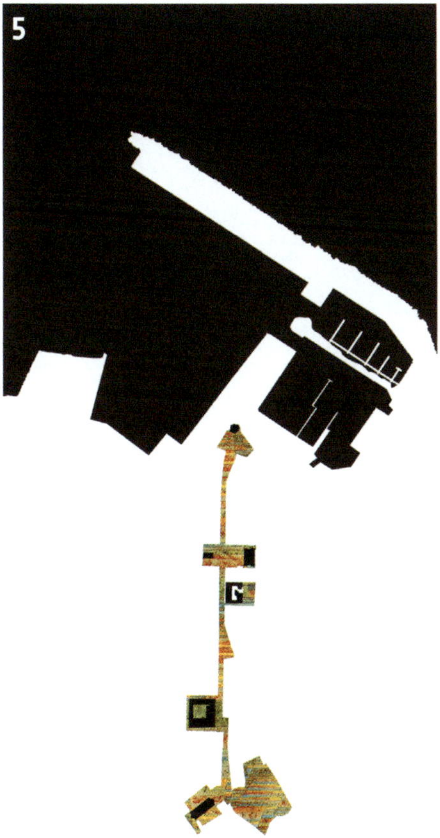

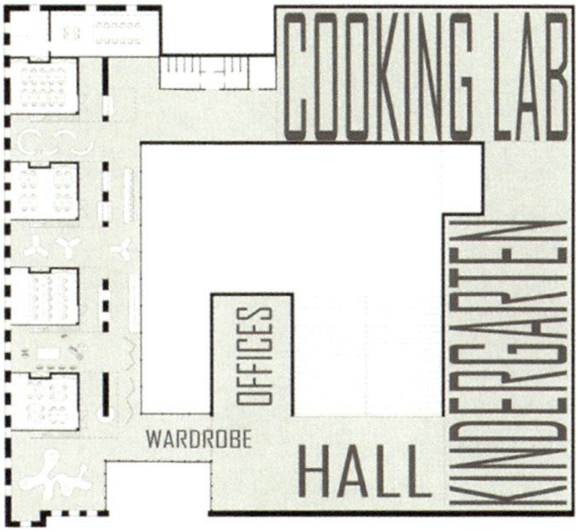

7

8

9

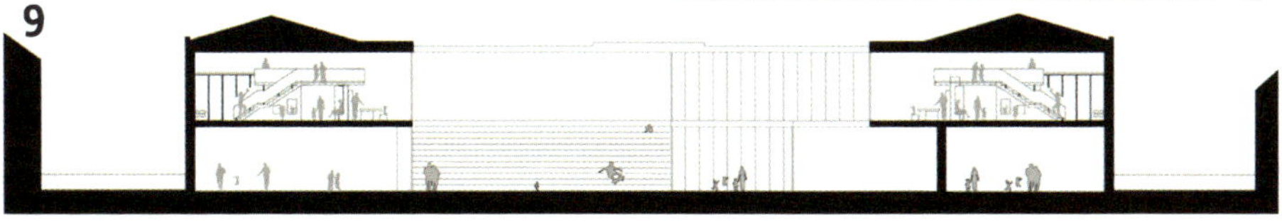

5.6

PT6

Playground and satellite district

Valter Caldana with Sabrina Scalas and Maria Letizia Angius, Francesca Camboni, Francesco Chiacchera, Salvatore Manca, Daniele Menghini, Chiara Rinaldoni

Type: Playground
Built: 1960
Condition: partly in disuse
Focus: the district is an ex-IACP construction located in a peripheral area of the town of Porto Torres, where scale relations between buildings and public space lose the urban dimension. The residual spaces, unfinished and not planned, are an opportunity to reconsider the relations between different city fragments, giving the district back spaces for children, for study, play and sport.
Project: conversion
Municipality of Porto Torres

The neighbourhood unit as a collective housing model
Sabrina Scalas

The several forms of aggregation for community life and rational organization of dwellings, as individual houses or collective housing, are based on a series of experimentations realized during the 19th Century and the first decades of the 20th Century.
The widespread industrialisation, developed between these two centuries, leads doctors, hygienists, engineers, etc. to question themselves about the ways to solve problems like the overcrowding of cities, and to improve living and sanitary conditions of the working class.
In a first phase, the workers' village, "a compact set of housing and general services (schools, bathrooms, wash houses, restaurants, clinics), is one of the few planned solutions." (Calabi, 2004)

Even the Garden Cities theorized by Ebenezer Howard in the early 1900s, start from the idea of reducing the congestion of the historic town thanks to the realization of new self-sufficient settlements, provided with housing and services.
One of the first experimentation of Howard's theories was realized by Barry Parker and Raymond Unwin through the construction of the cities of Letchworth and Welwyn, where dwellings were put in the middle of the sites and organized along curved and tree-lined streets.
The several European experimentations later found an evolution in the American urban planning policies with the development of the concept of neighbourhood units, "*theorized by Clarence Arthur Perry in 1929, in an attempt to define the idea of aggregation of dwelling able to strengthen community life.*" *(Montaner, 2008)*
The neighbourhood unit was an independent urban structure organized on a roadway consisting of main and secondary axes. In the centre of the settlement were localized all facilities for the development of collective life, like parks, playgrounds, shops and schools.
Perry's theories were experimented in the settlement of Radburn, New Jersey, in 1928 by Clarence Stein and Henry Wright with a team of architects of the "*Regional Planning Association of America*" founded by Clarence Perry and Lewis Mumford.
The theme of neighbourhood units will be utilized in the 50s by the various components of Team10 and experimented on several occasions as on the Central Carrières project in Casablanca by Georges Candilis and the GAMMA Group of 1952, or even on the Alexanderpolder project in Rotterdam proposed by Jacob Bakema and OPBOUW Group of 1953.
The satellite neighbourhood of Porto Torres, examined during the ILS workshop, can be assimilated to these international trials, both for reasons of its realization and for the urban layout that presents. Realized in the 60s and 70s of last century, in order to improve the living conditions and sanitation situation of the inhabitants of the ancient old town district built around the basilica of San Gavino, it presents an urban layout that follows a pinwheel neoplastic composition in which the various buildings are organized around the central space of the square.
Buildings of several sizes and typologies, with row houses and multi-storey blocks, define an urban space with more green areas. However, the space quality of the intervention is not connected with a social quality of community life. The neighbourhood lacks several public facilities such as schools, that are the basis of the theories of neighbourhood unit operation, that, as in the English elaboration of the Mars Group for the London plan of 1938, defined the minimum number of inhabitants of the neighbourhood unit depending on the numbers of children required to run a school and therefore the essential services related to the unit itself.
The ILS workshop wants to improve the neighbourhood human conditions, with a new space for learning, able to give a new centrality to this district of Porto Torres, i.e. a civic centre as place for social aggregation and landmark for all citizens that José Luis Sert defines as "*nucleus of urban culture*". (Sert, 1942)

References

- Donatella Calabi (2004), *Storia dell'urbanistica europea*. Milano, Mondadori
- Josep Maria Montaner (2008), *Sistemas arquitectónicos contemporáneos*, Gustavo Gili Editore, Barcelona
- José Luis Sert (1942), *Can Our Cities Survive? An ABC of Urban Problems, their Análisis, their Solutions*, The HUP, Cambridge

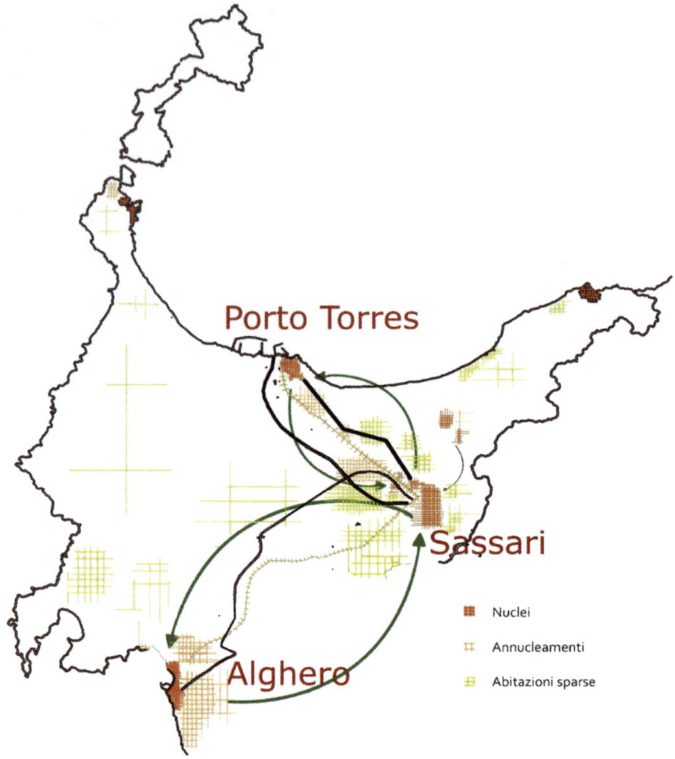
Porto Torres
Sassari
Alghero
Nuclei
Annucleamenti
Abitazioni sparse

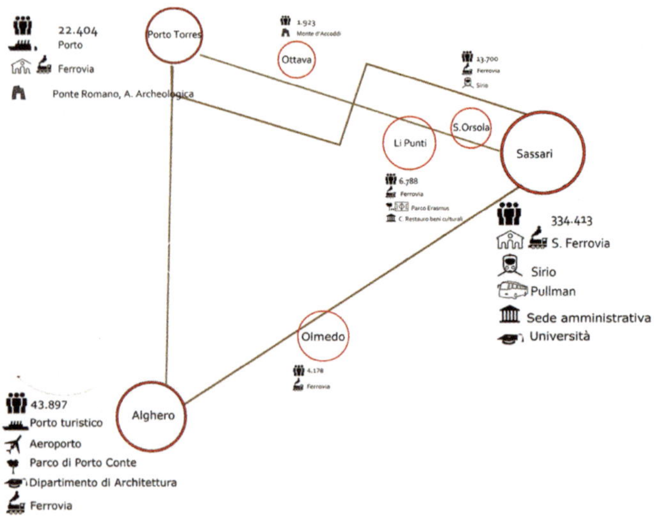
22.404
Porto
Ferrovia
Ponte Romano, A. Archeologica
Porto Torres
1.923
Monte d'Accoddi
Ottava
13.700
Ferrovia
Sirio
S.Orsola
Li Punti
Sassari
6.788
Ferrovia
Parco Erasmus
C. Restauro beni culturali
334.413
S. Ferrovia
Sirio
Pullman
Sede amministrativa
Università
Olmedo
4.178
Ferrovia
43.897
Porto turistico
Aeroporto
Parco di Porto Conte
Dipartimento di Architettura
Ferrovia
Alghero

LEGENDA

- Servizi sportivi
- Skatepark e campetto sport
- Aree pedonali
- Spazi interstiziali
- Aree carrabili

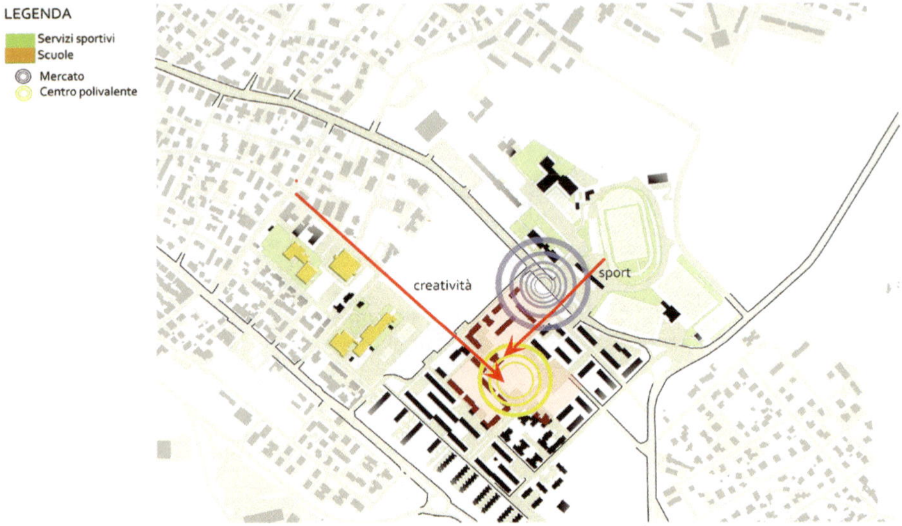
LEGENDA
Servizi sportivi
Scuole
Mercato
Centro polivalente
creatività
sport

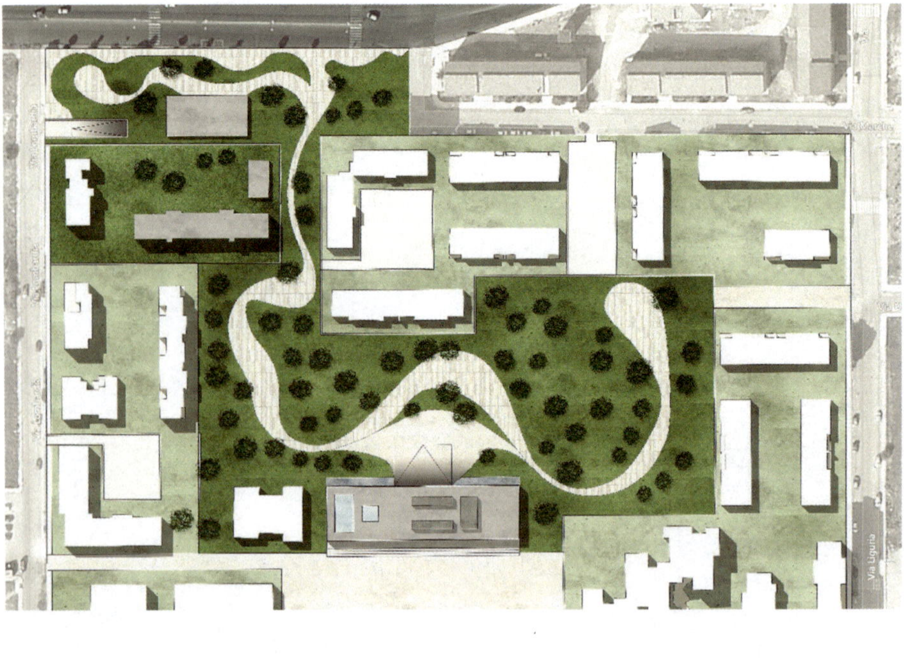
Via Liguria

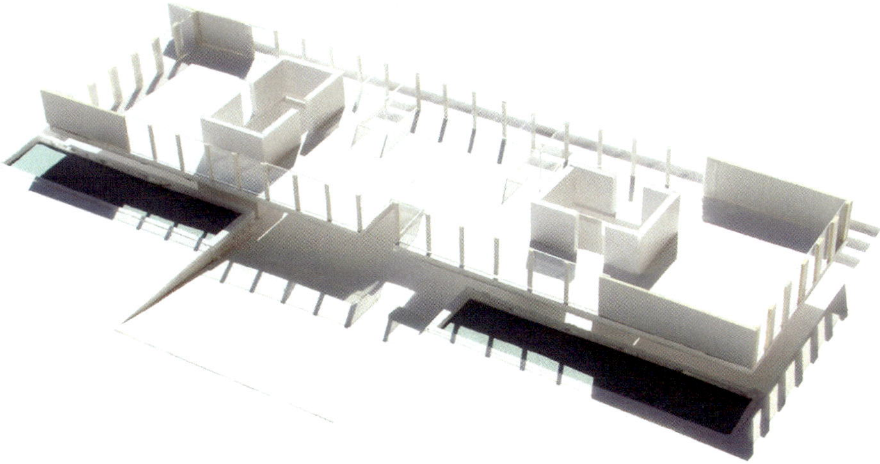

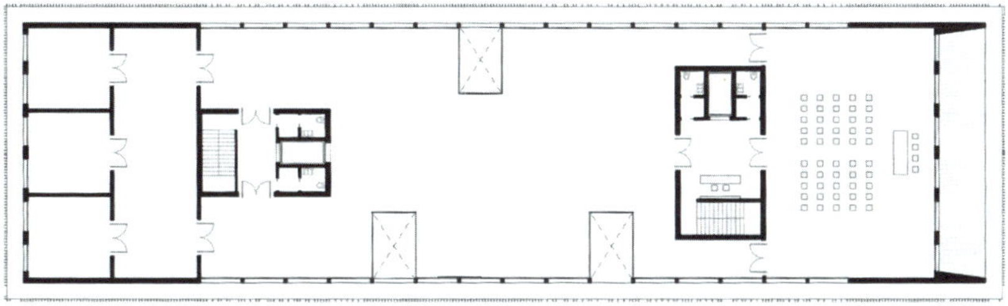

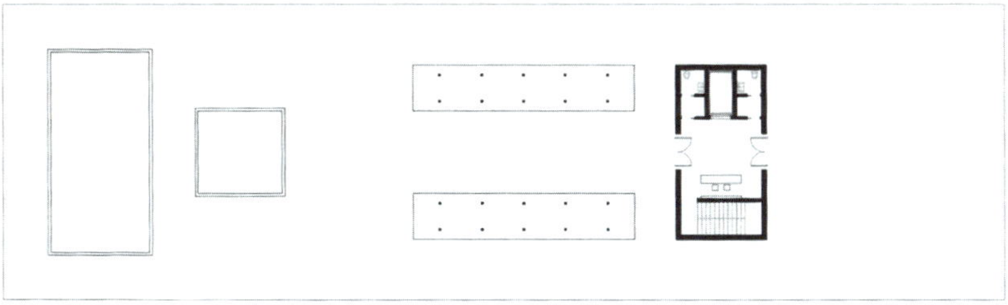

ILS -Innovative Learning Spaces
http://ils2016.wixsite.com/uniss

Scientific Director
Massimo Faiferri

Scientific Committee
Massimo Faiferri, Samanta Bartocci, Alessandra Lai, Augusto Marcelli, Fabrizio Pusceddu, Francesca Rango

Organizing Committee
Francesca Arras, Lino Cabras, Mario Casciu, Rosa Manca, Sabrina Scalas, Valentina Talu

Coordination of Work Groups
Valter Caldana (Università Mackenzie San Paolo)
Alfonso Femia e Simonetta Cenci (5+1 AA)
Massimo Ferrari (Politecnico di Milano)
Patrick Fransen (NOAHH)
Josep Mias (MIAS Architects)
Gianluigi Mondaini (Università delle Marche)
Joao Nunes, (PROAP)
Matteo Scagnol e Sandy Attia (Modus architects)

Teachers and Speakers
Fabio Bacchini
Antonello Marotta
Claudia Tinazzi
experts in pedagogy
Paolo Calidoni
Filippo Dettori
Donatella Petretto
Valentina Talu
Beate Weyland
experts in child neuropsychiatry
Alessandro Zuddas
experts in energy, materials and ICT
Giancarlo Della Ventura
Paolo Garau
Luca Malfatti
Augusto Marcelli
Giovanni Mazzitelli
Antonello Monsù

ILS
Innovative Learning Spaces

Edited by
Massimo Faiferri
Samanta Bartocci
Fabrizio Pusceddu

Published by
ListLab
info@listlab.eu
listlab.eu

Art Director & Production
Blacklist Creative, BCN
blacklist-creative.com

Editorial Director
Alessandro Martinelli

ISBN 9788899854393

Series BAB∃L

Printed and bound in the European Union
2020

On the cover Muri to listen
of Massimo Ferrari and Claudia Tinazzi

Sales, Marketing & Distribution
distribution@listlab.eu
listlab.eu/en/distributori-promotori-ditributors-promotors/

For more information concerning ListLab's Scientific Boards please visit the webpage:
listlab.eu/en/board-comitati-listlab/

ListLab was established in 2007 and has elaborated on the idea of an international editorial laboratory with a multidisciplinary approach to architecture, planning, arts, photography, and design. **List Group**, found in 2021, aims at creating networks and promoting debates and cultural exchange, but also organize events from which new knowledge about architecture, cities, and landscape can develop. Today, List Group is composed of **ListLab**, the publishing house, **Blacklist**, the graphic design studio, **Instaura**, the informational weblog, and **Us/Them/Yours**, a creative agency that aims at a multimedia approach to information.